THUNDERSTONES
AND SHOOTING STARS

Thunderstones
and Shooting Stars

THE MEANING OF METEORITES

Robert T. Dodd

HARVARD UNIVERSITY PRESS

Cambridge, Massachusetts, and London, England
1986

Library of Congress Cataloging-in-Publication Data

Dodd, Robert T., 1936–
    Thunderstones and shooting stars.

    Bibliography: p.
    Includes index.
    1. Meteorites.   I. Title.
QB755.D63   1986   523.5'1   86-7563
ISBN 0-674-89137-6 (alk. paper)

To William A. Caldwell (1906–1986), whose
daily column in the *Bergen Evening Record*
inspired millions of readers to think
and led at least one to write

Preface

Meteorites have captured the imagination of man throughout his long history on Earth. Fiery meteors or "shooting stars," the streaks of light that herald the arrival of extraterrestrial material in the Earth's atmosphere, appear in the art and literature of all cultures. Our forebears regarded them as omens and treated the meteorites (literally "thunderstones") that proceeded from some of them as sacred objects: one meteorite, Hopewell Mounds, was discovered in an Indian burial mound in Ohio; two others, Ogi and Sasagase, were found in Japanese temples. That many iron meteorites show evidence of forging—one, the New Mexico iron, is in the form of a hand ax—indicates that early man both revered such objects and put them to use. Clearly, our involvement with meteorites is of very long standing.

But why study them? What is there about meteorites that has caught the attention of a small army of astronomers, geologists, chemists, physicists, and even a few biologists? What is it that leads some men and women to brave the rigors of the Antarctic ice cap to recover meteorites, and makes others work through weekends and holidays to know them better?

One answer is that meteorites are the oldest rocks we have, relics of that dim past when a swirl of interstellar dust and gas gave birth to the sun and its family of planets and satellites, asteroids and comets. The great majority of meteorites formed before the Earth did. They record the first steps in the history of the solar system, steps of which all direct evidence has been erased from the Earth, the moon, and very probably the other planets as well.

A second answer is that meteorites come from space, an environment which we have visited many times in the last twenty years and in which we shall travel often in the future. They carry a record of that environment, in particular its radiation, that can help us plan for human occupation of space. It is true that we can also get such information from unmanned spacecraft, but meteorites bring it to us without cost. As one scientist put it, meteorites are "the poor man's space probe."

A third reason for scientific interest in meteorites is that they have played a role, however occasional, in the history of our own planet and its inhabitants. Huge meteorites have struck the Earth in the past, and there is considerable evidence that some of these objects intervened in the slow evolution of life on our planet. Although the jury is still out, it may be that one of the great mysteries of geology—the disappearance of the dinosaurs and many other life forms 65 million years ago—has a cosmic explanation. It is even possible that the origin of life itself is related in some way to the rich soup of organic compounds that we find in some kinds of meteorites.

We study meteorites, then, because of an inborn fascination with our beginnings: because they, and they alone among objects that we can hold in our hands and study in our laboratories, can carry us back to and perhaps even beyond the origin of the solar system.

This book summarizes what we know and think about meteorites. I wrote it out of a sense of obligation to those who have asked why, with an Earth full of rocks to study, this geologist has chosen to spend the last two decades studying meteorites.

Many people have contributed to this book. Wherever possible, I have acknowledged their generosity in the text or in figure captions. I am particularly grateful to Lois Koh, who drafted most of the diagrams; to Michael Lipschutz and Roy S. Clarke, Jr., who critically reviewed the manuscript; and to my editors, Kenneth I. Werner and Mary Ellen Geer, the former for seeing the book through its early development and the latter for helping me bring it to completion. Finally, I am grateful to my wife, Marya R. Dodd, whose support of the project included many long hours of proofreading.

Contents

Meteor Crater, Arizona, aerial view. The crater is 4,100 feet wide and 600 feet deep.

(Photograph courtesy of the Department of Library Sciences, American Museum of Natural History; negative no. 331534.)

1
Target
Earth

A tiny light blazes high in the night sky, then sputters and goes out as the speck of dust that caused it falls softly to Earth. Lost among countless other grains that cover our planet's gritty surface, the tiny meteorite will pass unnoticed and go unnamed, but it and others like it will add hundreds of tons of new material to the Earth in a single day.

In the state of Chihuahua, northern Mexico, the stillness of a winter night is shattered by a fireball that hurtles across the sky, flies apart with thunderous detonations, and scatters thousands of stones over an area 30 miles long and several miles wide. The timing of this meteorite fall is perfect: it is February 1969, and scientists all over the world are poised in wait for the first lunar samples, which will arrive six months later. The new meteorite, named Allende for its place of fall, will quickly find its way to dozens of laboratories around the world, where it will play a lively overture to the flight of Apollo 11 and go on to rewrite the first chapters of the history of the solar system.

In the dry plateau country of northern Arizona, a huge mass of nickel-iron plunges into the ground and explodes, hurling bits of itself and its sandstone target for miles in every direction and leaving a hole 4,100 feet wide and nearly 600 feet deep (see frontispiece). Some 20,000 years later, men will try in vain to mine the meteorite for its iron and will argue about the meaning of the tiny diamonds they find within it. But in the end, the Canyon Diablo meteorite will be much less famous than its scar: an impact structure so perfect and so well preserved that astronauts will study it as they prepare to explore

1

similar features on the moon, a structure so typical that it needs no name more specific than Meteor Crater.

Bits of extraterrestrial debris continually bombard the Earth as it makes its way around the sun. As the previous examples show, these objects, which we call *meteoroids,* come in a wide variety of sizes and kinds, from tiny masses of fluffy dust like the first example to 2-ton chunks of rock like Allende and 100,000-ton pieces of metal like Canyon Diablo. What are these objects? Where do they come from? Where, when, and how did they form? What do they tell us about the history of the solar system and, in particular, the Earth?

These questions will occupy our attention throughout the book. First, though, a few basic questions need to be answered about meteoritic material. How much of it reaches the Earth? How common are meteorites that are large enough to recover and study? How common and how dangerous are crater-forming meteorites?

The Atmospheric Filter

Before a meteoroid can reach the Earth to become a meteorite, it must pass through the atmosphere. How successfully it does so determines what it is like when it lands and how useful it will be to meteoriticists.

All meteoroids enter the Earth's atmosphere at speeds of 7 to 19 miles per second. The lower limit is the Earth's escape velocity: the speed that an object leaving the Earth's surface must achieve to break away from our planet's gravitational field. The upper limit, really an estimate, is the speed beyond which an incoming object should be destroyed on impact with the atmosphere. (If the concept of impact on a fluid like air is troublesome, think back to the last time you dived into a swimming pool and landed flat: at high impact speeds, fluids behave remarkably like solids!)

When meteoroids enter the atmosphere at such high speeds, all but the tiniest of them are heated enough by friction to melt and glow, producing the streaks of light that we call shooting stars or *meteors.* The smallest meteoroids escape this fate because their surfaces are so large in comparison to their mass that they radiate frictional heat as fast as it develops and thus avoid melting. Such objects pass through the atmosphere nearly unchanged to reach the Earth's surface as a subtle rain of *cosmic dust.*

All meteoroids with diameters larger than about one ten-thousandth of an inch experience melting on their way through the atmosphere. The smallest melt completely, and the tiny spherules that result fall to Earth as *micrometeorites,* like the one described at the beginning of this chapter. Micrometeorites and similar ablation droplets (Figure 1.1) retain some of the chemical properties of the meteoroids from which they came, but melting completely erases their original mineralogy and texture. This fact makes it very difficult to work out their preterrestrial history.

Meteoroids that weigh more than a gram (about four-hundredths of an ounce) undergo only surface melting. As such objects pass through the atmosphere, air resistance strips away the molten material as it forms, carrying heat away from the meteoroid and protecting its interior in much the same way that an ablating heat shield protects

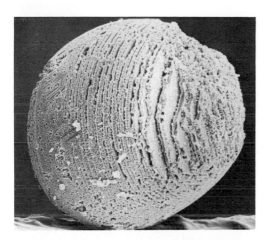

Figure 1.1 Scanning electron microscope photograph of a micrometeorite, or, more probably, an ablation droplet. The object, about 1/32 inch in diameter, consists mainly of plates of olivine in parallel arrangement. (Photograph courtesy of Professor Donald Brownlee.)

Figure 1.2 The Miller, Arkansas, meteorite, showing its thin, black, shiny fusion crust and flow lines. The specimen is approximately 3 inches in diameter. (Photograph by Thane Bierwert, courtesy of the Department of Library Sciences, American Museum of Natural History; negative no. 411569.)

the crew inside a space craft. One-gram and larger meteoroids acquire a glazed surface, or *fusion crust* (Figure 1.2), which is a characteristic feature of meteorites and helps to distinguish them from terrestrial rocks. Beneath their fusion crusts, these meteorites remain much as they were in space. Because such objects are large enough to be recovered individually, I shall call them *recoverable meteorites* here to distinguish them from micrometeorites.

In addition to heating incoming meteoroids, atmospheric transit slows them down. All but the biggest meteoroids strike the Earth's surface at a small fraction of their preatmospheric speed—typically 200 to 400 miles per hour—with the result that we find most newly fallen meteorites either on the ground surface or in holes that are only slightly larger than themselves. Only meteoroids that weigh more than about 350 tons reach the Earth at close to their original speed, entering the ground and exploding to produce craters much larger

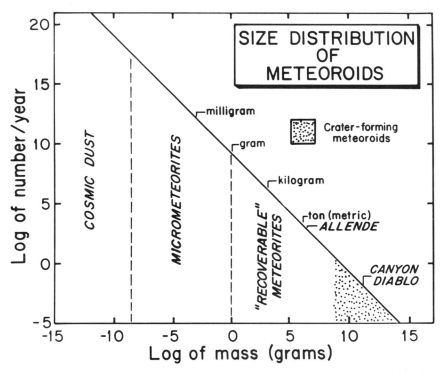

Figure 1.3 Relationship between mass and frequency for meteoroids in the vicinity of the Earth. The mass ranges for various kinds of meteorites are indicated. Logarithmic scales are used to accommodate very large ranges for both variables (e.g., on the mass scale, 0 = 1 gram, 1 = 10 grams, 2 = 100 grams, and so forth; on the frequency scale, 5 = 100,000, −5 = 1/100,000).

than themselves. Obviously, Canyon Diablo was such a *crater-forming meteorite:* Meteor Crater is about 50 times as wide as the chunk of nickel-iron that formed it.

In terms of size, then, we can distinguish four kinds of meteoritic material: cosmic dust particles, whose small sizes permit them to pass through the atmosphere intact; micrometeorites, which include small, wholly melted objects and droplets of material ablated from the surfaces of larger meteoroids; recoverable meteorites, the interiors of which are unchanged beneath a thin fusion crust; and crater-formers, which are recoverable meteorites whose large size permits them to strike the Earth hard enough to explode after impact. Figure 1.3 shows the approximate mass limits for these four types of material and the frequency with which objects of different sizes strike the Earth.

Weighing the Crop

How much extraterrestrial material strikes the Earth today? How much has struck it in the past? These questions are important for both practical and scientific reasons. Such material is a potential hazard for spacecraft operating near the Earth and, conceivably, for life on our planet. It also adds mass to the Earth and might, if abundant enough, change its composition. Finally, the amount of incoming extraterrestrial material concerns meteorite researchers, whose livelihood depends on recovering it for study.

There are many ways to estimate the flux of meteoritic material. The simplest approach—just counting the meteorites that fall in a given area and dividing by the period of observation—turns out to be the least reliable, because we find only a tiny fraction of potentially recoverable meteorites. For this reason, meteoriticists prefer to base flux estimates on the much more numerous small objects—cosmic dust and micrometeorites—that strike the Earth. The methods they use include counting or collecting micrometeorites in the atmosphere, collecting them from terrestrial sediments or snow, and measuring the flux of meteors, the fiery harbingers of meteorites.

Atmospheric collections of micrometeorites have been made both at the Earth's surface and far above it. Surface collection is easiest—a few years ago, one scientist just laid strips of flypaper on a rooftop—but the harvest of extraterrestrial material obtained in this way is heavily contaminated by sedimentary, volcanic, and industrial particles. Airborne collection, using airplanes, rockets, and artificial satellites, avoids the problem of contamination, but unfortunately, the fluxes measured in this way have proved to be variable and hard to interpret.

Sampling deep-sea sediments has been a popular means of flux estimation in the hundred or so years since Murray and Renard recovered black, magnetic "cosmic" spherules from deep-sea clays collected by the *Challenger* expedition in 1876. Such clays have an important advantage over other sediments because they form very slowly and far enough from land to minimize dilution by sand and silt from the continents. Industrial contamination is also minor and restricted to the topmost layers of the clay. (Polar snow—also a sediment, though not usually considered as such—offers similar advantages.)

Particle-by-particle analysis of micrometeorites recovered from sediments can tell us not only how much material is hitting the Earth but what kinds. However, both recovery and analysis of individual spherules are hard work, and researchers have therefore sought faster means to estimate the flux. One such method makes use of the fact that some chemical elements—iridium, for example—are much more abundant in meteorites than they are in common rocks of the Earth's crust. If we know the iridium content of typical meteoritic material, measurements of the iridium content of a sediment and the time interval during which the sediment was deposited can be used to calculate the flux of meteoritic material during deposition.

A similar approach takes advantage of the presence in meteorites of short-lived radioisotopes that were formed by cosmic rays while the meteorites were in space. One such isotope is aluminum-26. Since it changes, or decays, to magnesium-26 very rapidly in geological terms—half is gone in 740,000 years—we can be certain that any aluminum-26 that was present in the Earth when it formed has long since vanished. Thus any aluminum-26 that we detect in a sediment signifies extraterrestrial material, and the amount present can be used to calculate the abundance of such material. Once again, if we know how long it took for the sediment to form, we can easily calculate the flux of meteoritic material during that time period.

Both trace elements and short-lived, cosmic ray–induced radioisotopes are excellent tracers for meteoritic material, but the former have one important advantage over the latter. Aluminum-26 decays so fast that we can only detect it in modern sediments and very young sedimentary rocks. In contrast, we can measure iridium and other meteorite-related elements in ancient rocks as well and can therefore estimate the meteorite flux for earlier periods in geological history. As we shall see in Chapter 11, unusually high iridium contents in marine clays deposited at the end of the Cretaceous period, about 65 million years ago, have led to the suggestion that a huge meteorite fell at that time and was responsible for sweeping changes in life on Earth, including the demise of the dinosaurs.

All the methods of flux estimation discussed thus far measure material that has reached the Earth. In addition, however, we can estimate the meteoroid flux by observing the frequency of meteors. Meteor counting, using both meteors that are bright enough to darken a photographic plate and smaller ones that must be detected by radio techniques, carries two bonuses: it tells us not only how much material is entering the Earth's atmosphere, but also the size distribution of that material and the directions from which it comes.

In view of the wide variety of methods used to estimate the amount of incoming meteoritic material, it is not surprising that flux estimates vary widely. Fortunately, most of the recent estimates lie in a narrow range, between 100 and 1,000 tons per day for the entire Earth. The true flux probably lies somewhere between these limits. This amount of material would obviously seem immense if it were dumped in your backyard, but it is actually tiny when spread over the Earth's vast surface. If the same flux persisted throughout our planet's 4.55-billion-year history, it would yield a layer of meteoritic material about 5 feet thick—little enough to be lost among the products of the Earth's own processes. Thus meteorite infall now has no measurable effect on the composition of the Earth's crust. Studies of micro-meteorites in ancient salt deposits suggest that this has been true for the last half-billion years, and the frequency of impact craters on the lunar surface suggests that the meteoroid flux has been more or less constant there for perhaps 3.0 to 3.5 billion years. However, the battered lunar highlands testify to much higher meteoroid fluxes during our satellite's first billion years. No doubt the Earth also underwent a cosmic pounding early in its history, but all evidence of this has been erased by aeons of erosion, volcanism, and mountain building.

Meteorite Sizes:
Hazards and Opportunities

With a good estimate of the amount of meteoritic material that strikes the Earth, the next question is how this material is distributed by size. How much arrives in pieces that we can hope to recover for study? How common are inconveniently large objects—meteoroids that are big enough to be destructive?

Studies of meteors tell us that small meteoroids are much more abundant than large ones. In fact, a tenfold decrease in number accompanies each tenfold increase in weight, a relationship that is shown by the diagonal line in Figure 1.3. That this relationship accurately describes the size distribution of material approaching the Earth is confirmed by size data for large meteorites and for the minor planets or asteroids, which are important sources of meteorites.

With the help of Figure 1.3, we can attempt to answer the questions posed above. First, if we take a meteorite that weighs 1 gram as the smallest that we can hope to recover, we find that about 1.6 *trillion* objects that big or bigger strike the Earth each year, or about eight meteorites per square mile of our planet's surface. If we could recover all of these meteorites, our collections would grow at a staggering pace! Unfortunately, as we shall see, we recover very few 1-gram meteorites, and there is a wide gap between "recoverable" and "recovered" for much larger objects.

To evaluate the risk of being struck by a large meteorite, we need more exact definitions of "risk" and "large." Since a baseball traveling at 90 miles per hour can fell a batter, it is clear that a 2-pound meteorite moving twice to four times that fast can do a lot of damage, and we can expect about a million such objects to hit the Earth each year. Fortunately, this number translates to a very low probability for a particular spot on Earth: though meteorites have struck buildings—one went through the roof of a house in Wethersfield, Connecticut, in November 1982—only one meteorite-related injury has been reported, and there have been no known fatalities.

Crater-forming meteoroids—those that weigh at least 350 tons—can, of course, do much more damage, since their high speeds give them more energy than the same weight of TNT. Figure 1.3 tells us that about five such meteoroids strike the Earth each year. A meteoroid at least as big as the one that created Meteor Crater—roughly 100,000 tons—should come along every 60 years or so. These figures seem high enough to justify some concern for public safety, but two factors work to lower the risk that they imply. First, most such meteorites fall in the oceans, which cover about three-fourths of the Earth's surface. Second, as the description of Allende's fall illustrates, large stony meteorites are brittle. Most of them break up in the atmosphere and land as many small fragments rather than a single mass. It is likely that most terrestrial meteorite craters, though certainly not all, were made by the less brittle meteorites composed largely of metallic iron, which are far less common than stony meteorites.

To get a more realistic view of the likelihood of a catastrophic meteorite impact, we have to correct the fluxes read from Figure 1.3. Specifically, we must divide them by 4 (to obtain the flux for dry land) and then by 20 (to get the dry-land flux for irons and stony-irons). When we do so, we find that a crater-forming meteoroid should hit the Earth every 16 years or so, and we can expect objects as big as Canyon Diablo to hit it at intervals of about 5,000 years. Although these estimates are by no means exact, an insurance actuary would find them comforting. They show that destructive meteorite falls are rare events relative to the time scale of human history. When we consider that most of the Earth's land surface is thinly populated or uninhabited, it becomes clear that—novels and video games notwithstanding—meteorite impact should rank very far down the list of human concerns.

An actuary might, on the other hand, be reluctant to write a "meteorite policy" for the Earth, since on the vast time scale of geological history, catastrophic impacts become possible and indeed likely. Figure 1.3 suggests that a 1.1-billion-ton iron or stony-iron meteoroid should strike dry land every 50 million years or so. As immense as such an object sounds, it would be only slightly more than 2,000 feet in diameter—smaller than the tiniest visible asteroid. Clearly even larger impacts are possible, and it is by no means frivolous to speculate that they played a role in our planet's geological history.

2

Meteorite Recovery

Although trillions of potentially recoverable meteorites have fallen in the 180 years since scientists first recognized them as important objects, we have samples of fewer than 2,500 falls. This sharp contrast between a bountiful crop and a meager harvest underscores an important difference between meteoritics and the other natural sciences. A geologist who is interested in a particular kind of rock or a botanist concerned with a specific plant can usually go somewhere on Earth to collect it, but meteorites can fall anywhere and at any time. Hence meteoriticists have to trust that someone will see and report a meteorite fall or will recognize a meteorite if it turns up in his garden.

That meteorite recovery is a chancy business explains why many laymen pursue meteoritics as a hobby, and some even maintain their own meteorite collections. Meteoritics is the rare science in which an amateur can make an important contribution. Few nonspecialists can hope to name a new species of bird, and none will discover a new chemical element, but men and women of every walk of life have found meteorites or, by reporting their flight paths, have helped others find them. It is not surprising that the Meteoritical Society was founded by and still includes many amateurs.

When a meteorite falls, meteoriticists try to recover it as quickly as possible. One reason for this is that we rely on eyewitness descriptions of a meteoroid's line of flight to tell us where to look for a meteorite, and such descriptions—like the testimony of witnesses to a crime—become less and less reliable as the observers compare notes. Accuracy is very important, for even a tiny difference in the flight path—whether an object passed over the big oak tree or the maple next to it—can shift the search area by miles.

A second reason for haste is that meteorites begin to corrode, or weather, as soon as they land. They contain nickel-iron, iron sulfide, and other minerals that can survive indefinitely in airless space but react quickly with the Earth's moist, oxygen-rich atmosphere. Like a hoe left in the garden over the winter, meteorites rust and their original character is soon lost.

A third reason concerns the cosmic ray–induced radioisotopes mentioned in the previous chapter in connection with flux estimates. Production of these isotopes ceases as soon as a meteorite enters the Earth's shielding atmosphere, and they start to decay. Some isotopes, aluminum 26 for example, decay slowly enough for leisurely study, but others have half-lives of days or hours. Thus a meteorite that is recovered just after its fall carries a precious but elusive record of its radiation experience, and it gets priority treatment. A 4-pound chondrite that fell in a Connecticut living room on the night of November 8, 1982, was on its way to a nuclear counting laboratory in the state of Washington just two days later.

In view of the great value of fresh meteorites, or *falls,* it is unfortunate that our collections include only about 800 such objects, most of them samples of a few common types. If we hope to learn anything about the more exotic varieties of meteorites, we must also study *finds:* meteorites that are recovered, usually by accident, some time after their fall. Unfortunately, finds vary widely in quality: some are nearly fresh and we can use them for all but the most exacting chemical and biological studies; many are rusty bits of rock or metal that are hardly worth carrying home.

One important goal of meteoriticists is to improve our ability to recover fresh falls. Another is to locate good sources of finds: places on Earth whose climate favors preservation of meteorites and whose terrain makes them easy to find. Progress toward the first goal has been modest and is likely to remain so. On the other hand, the last decade has witnessed the opening of an immense trove of well-preserved finds that has increased our supply of meteorites dramatically and has given meteorite research a new dimension.

Falls

Studies of meteor trajectories tell us that the large, dense meteoroids that yield meteorites fall at about the same rate over most of the Earth, with only a slight decrease near the North and South Poles.

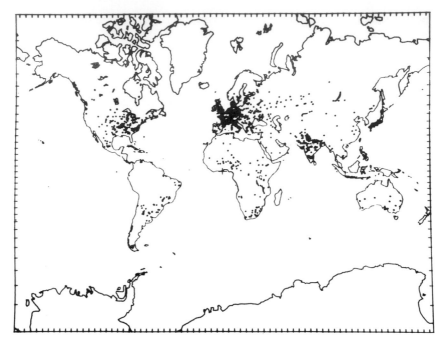

Figure 2.1 Geographic distribution of meteorites recovered after observed falls. (From Hughes, 1981, reproduced by permission of the author and *Meteoritics*.)

Since a meteor must be seen before we can recover its meteorite, we might expect the likelihood of recovery to be greatest in areas of high population density. We might also expect a meteorite's size to affect the chances for recovery, for once eyewitness descriptions of a fall have shown us where to look for a meteorite, we must still pick it out from innumerable terrestrial rocks. Few meteorites have the good grace to land on a road, as the Lost City , Oklahoma, chondrite did in 1970.

As Figure 2.1 shows, population has the anticipated effect on meteorite recovery.* Most falls have been recovered in the Earth's mid-latitudes and in densely populated areas within that belt. The crowded European countries and Japan have high recovery rates; thinly populated countries like the United States, the Soviet Union,

---

*The sources of the data in Figures 2.1 and 2.2 are the *Catalogue of Meteorites* (Hey, 1966; Hutchison et al., 1977). Recently updated (Graham et al., 1985), the *Catalogue* lists all known non-Antarctic meteorites with information on their recovery, properties, and classification. It is an invaluable resource.

Canada, and Mexico have modest ones (Figure 2.2). We can also see population effects for different regions within countries. For example, the crowded northeastern corridor of the United States—roughly between Boston and Washington, D.C.—has a far better record than the rest of the country, and the Ukraine is far ahead of the rest of the USSR.

If we look at Figure 2.2 more carefully, though, we can see that factors other than population must be at work too. Why are the French particularly good at recovering falls? Why do India and crowded Japan rank well below Germany and the United Kingdom? The reasons for these peculiarities are historical. French scientists were among the first to grasp the significance of meteorites, just after 1800. This fact and the presence of a dispersed, largely rural population explain why the French were able to recover 27 falls between 1800 and 1850—a record that no other nation has approached.

If the French came to meteoritics early, the Indians and Japanese arrived late (Figure 2.3). In both cases, a surge in efficiency of recovery coincided with the opening of the country to Western influence. It appears that heightened interest in meteorites was one of the happier

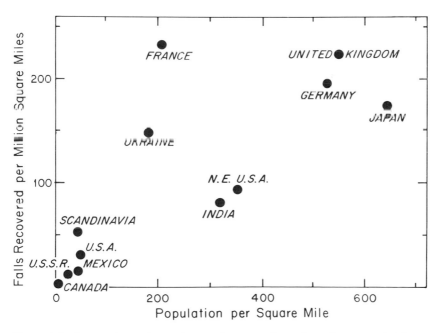

Figure 2.2 Comparison of population density and number of recovered falls for several nations and regions.

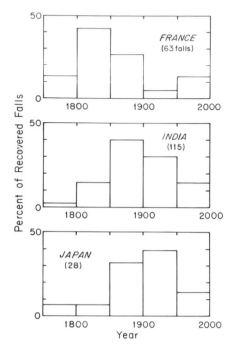

Figure 2.3 History of fall recovery for France, India, and Japan since 1800. Data for the years 1950 to 2000 are extrapolated from those for 1950–1975.

results of Commodore Perry's visit to Japan in 1853 and the beginning of British rule in India in 1858.

At the risk of overstretching the data, we can note two more implications of Figure 2.3. It is evident that the otherwise diligent French collected few falls between 1900 and 1950, quite likely because they were engaged in two disastrous wars during that period. Germany's record also sags during the same period, though less strikingly.

A surprising implication of Figure 2.3 is that we have recovered relatively few falls since World War II. It is possible that the meteoroid flux has waned, but it is more likely that our data, which are based on the period from 1950 to 1975, are simply incomplete. Even the diligent keepers of the *Catalogue of Meteorites* are always somewhat behind in their tally. It will be interesting to look back a few years from now and see whether man's outreach toward the planets, which began about 1960, was reflected in a higher than normal recovery rate for meteorite falls.

As successful as the French were at meteorite recovery between 1800 and 1850, they collected fewer than one ten-thousandth of 1 percent of the potentially recoverable meteorites that fell in France during that period. At his best, man is a most inefficient meteorite collector. The reason for this huge discrepancy lies in the word "recoverable," for the likelihood that we will find a meteorite drops sharply with decreasing size. A 1-gram meteorite is, in principle, recoverable, but we find objects that small only rarely and under unusually favorable circumstances. The 1-gram Revelstoke chondrite and the 142-milligram Vilma chondrite, which fell in western Canada in 1965 and 1967, respectively, were easy to find because they landed on snow. Revelstoke and Vilma are two of only five falls in our collections that weigh less than 10 grams.

That such objects are seldom recovered is not really surprising, since a 10-gram stony meteorite is only as bit as a hazelnut. For that matter, a 100-gram stone is about the size of a golf ball, and a 1000-gram stone (about 2 pounds) is as big as a baseball. Even the last of these objects would be lost in my Long Island garden, which contains many head-sized boulders through the courtesy of a passing glacier.

Consideration of the sizes of meteorites does much to restore the luster of the French performance between 1800 and 1850, and it raises an interesting question: how big must a meteorite be to give us a good chance of recovery? This depends, of course, on what else is on the ground where it lands—a 2-pound meteorite would stand out on a stone-free lawn or in a well-tended carrot bed—but a comparison of the size distribution of recovered falls with that of meteoroids suggests that we recover almost every observed fall that weighs more than about 100 kilograms (220 pounds). We recover perhaps half of those that weigh 20 pounds, and about 10 percent of the 2-pounders.

It is clear, then, that an attempt to locate and recover a freshly fallen meteorite is a game played against long odds. Is there any way to improve those odds? We cannot do much about the size problem—it will always be hard to spot a meteorite in an area littered with terrestrial boulders—but we can try to narrow the search area by using cameras rather than human eyes to track fireballs, the brilliant meteors that drop meteorites. The first meteorite that was recovered with the help of a photographically determined trajectory is the Příbram chondrite, whose fireball was photographed in 1959 by two stations in a Czechoslovakian camera network that had been established to monitor man-made satellites.

Partly in response to the Příbram experience, researchers at the Smithsonian Institution's Astrophysical Observatory built the Prairie Network, an array of 16 automatic cameras that monitored the sky over a 300,000-square-mile area of the Midwest centered on Nebraska. They chose this area because its deep soil, rendered almost free of boulders by several generations of farmers, strongly favors recovery of any meteorite that falls there. As an aid to meteorite recovery, the now-defunct Prairie Network was a disappointment. Although its cameras recorded many fireballs, only one meteorite— Lost City, Oklahoma (1970)—was recovered. Other camera arrays have had similar histories: a Canadian network has also yielded just one meteorite (Innisfree, Alberta, 1977); a Czech-German network and one in the United Kingdom have yielded none. Fortunately, meteorite recovery is just one purpose of these camera networks, which also provide a wealth of data on the sizes and orbits of meteoroids.

We can improve the grim odds against recovery of fresh falls, but only slightly and at great cost. Thus major falls will always be important events, and meteoriticists will greet them, as we greeted Allende, with all the enthusiasm of prospectors galvanized by a rumor of gold.

Finds

Like falls, most meteorite finds come from the populous mid-latitudes (Figure 2.4). In the tropics, a sparse population conspires with a hot,

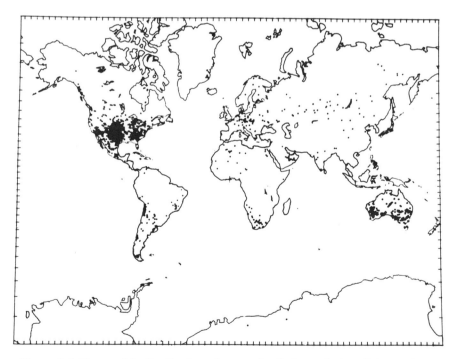

Figure 2.4 Geographic distribution of meteorite finds, exclusive of Antarctica. (From Hughes, 1981, reproduced by permission of the author and *Meteoritics*.)

Table 2.1 The eleven most meteorite-productive states

State	Population (per square mile)	Falls (per million square miles)	Finds (per million square miles)
Kansas	26	85	1,058
Colorado	17	29	566
Texas	36	34	479
Kentucky	75	99	470
New Mexico	8	25	436
Tennessee	84	71	426
Nebraska	18	26	414
Georgia	67	51	340
North Carolina	86	171	398
Oklahoma	33	72	272
Arizona	11	9	237

Source: Data are from the *Catalogue of Meteorites* (Hey, 1966; Hutchison, Bevan, and Hall, 1977).

moist climate that quickly destroys all kinds of rocks, meteorites included. Low returns from the polar regions are, however, solely a result of underpopulation, for meteorites are there and in very great abundance.

Population density explains the broad pattern in Figure 2.4, but some aspects of the distribution of finds require other explanations. For example, the United States, whose record of fall recovery is undistinguished (see Figures 2.1 and 2.2) leads all other nations in meteorite finds, both in absolute number (572 as of 1975) and in number per million square miles (158). Germany is a distant second with 73 finds per million square miles, Japan third with 56, and Mexico fourth with 55. France, which leads the world in falls, is fifth in finds, with only 14 per million square miles.

To some extent, our success in finding meteorites is related in a negative way to our success in recovering falls: the more meteorites we recover just after they land, the fewer we can find later. This may explain France's poor find record and that of the United Kingdom, which claims 21 falls but no finds. The great success enjoyed by the United States, however, demands other explanations. They emerge when we look at where most American finds have come from (Table 2.1). With few exceptions, the states that have yielded the most finds are those of the prairie and the arid southwest—the former because of deep, intensively cultivated, and thus largely boulder-free soil, and the latter because of a dry climate in which meteorites tend to weather slowly. By contrast, the populous Northeast has a moist climate, and its soil, pushed about by continental glaciers as recently as 10,000 years ago, is full of native boulders that frustrate our attempts to find meteorites. It is thus not surprising that crowded, busy New York has yielded few finds and New Jersey and Connecticut have yielded none.

But the American Midwest and Southwest have enjoyed another advantage as well: the active interest of many meteorite collectors, outstanding among whom is Harvey H. Nininger. Since the 1920s Nininger has traveled the country tirelessly, looking for meteorites himself and teaching farmers to look for them. In person and through his books, Nininger stimulated popular interest in meteorites long before they became scientifically fashionable. Nature made middle America rich in meteorites, but Nininger and others like him brought the harvest home.

Other areas—South Africa, coastal Australia, parts of India—share many of the advantages of the American prairie and Southwest, and it is not hard to see why these places have yielded many finds (see Figure 2.4). A concentration in Chile is more puzzling until we find that many of the iron meteorites found there are from a few large falls and that they were collected with particular diligence. A band of finds across central Asia also represents a special circumstance: the presence of the Trans-Siberian Railway, whose right-of-way marks a bulge in an otherwise sparse population.

Some areas of Figure 2.4 are surprisingly meteorite-poor. For example, although Japan shows many finds, most of nearby China is blank, despite a huge and largely rural population. The reason for this anomaly is political: until the early 1970s Western scientists knew almost nothing about meteorites in the People's Republic. We still know little, but a lively scientific exchange has begun. With any luck, Figure 2.4 will be the last such map to suggest that most of China is meteorite-free.

It is clear, then, that preservation and the absence of native boulders are the most important considerations when we go out to hunt for meteorites. My backyard in New York—and most likely yours—fails on both counts. There may be meteorites among the innumerable rocks in my garden, but they would be hard to find and too badly weathered to be worth the effort. A far better place to look for meteorites is a dry, flat, boulder-free area like Australia's Nullarbor Plain, where the rabbit-hunters can—with only slight exaggeration—spot meteorites with binoculars and scoop them up from moving motorbikes.

But even on the Nullarbor Plain, it is a long drive between meteorites. Is there any place on Earth where nature has both preserved meteorites and brought them together to make collecting easy? It turns out that there is: its only—though considerable—drawbacks are that it is hard to get to and very cold.

Antarctic Meteorites

The 1966 edition of the *Catalogue of Meteorites* lists only four meteorites found on the continent of Antarctica. One, the Adelie Land chondrite, was found by an Australian polar expedition in 1912 (Figure 2.5); the others, two irons and a pallasite, were discovered by similar expeditions between 1961 and 1964. The 1977 Appendix to the *Catalogue* lists several more meteorites that Japanese glaciologists collected in the Yamato Mountains of Enderby Land in 1969, and the 1985 edition raises the count to 43.

Future editions of the *Catalogue* will list hundreds, perhaps thousands, of Antarctic meteorites. Formerly collected from the ice cap by men on other missions, meteorites are now the object of special searches in Antarctica, and the results are dazzling: about 7,000 samples have been recovered to date, most of them from two localities with a total area of about 200 square miles. Since one

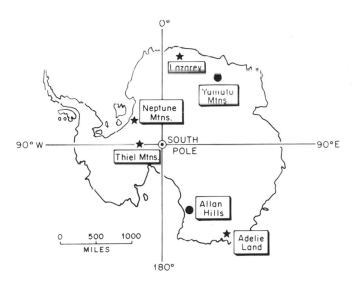

Figure 2.5 The continent of Antarctica, showing the locations of individual meteorite finds (*stars*) and major areas of meteorite concentration (*circles*).

meteoroid can yield many meteorite samples, this collection probably represents between 70 and 700 separate falls—a concentration of meteoritic material that puts all other collecting areas quite in the shade.

At first glance, Antarctica seems a poor place to look for meteorites. Because most of the heavy meteoroids that yield meteorites travel in the same plane as the Earth and the other planets (that is, the ecliptic plane), meteorite falls are slightly less frequent in the Earth's polar regions than at lower latitudes. Why, then, do tremendous concentrations of meteorites occur in the Yamato Mountains (see Figure 2.5), in the Allan Hills of Victoria Land, and in other places on the Antarctic continent? And what, other than their great abundance, makes Antarctic meteorites especially appealing to meteoriticists?

To answer the first question, we need to know something about glaciers. A glacier forms when snow accumulates year after year instead of melting entirely each spring. As the snow mass thickens, pressure converts the snow to granular material called firn and finally to solid ice. When the mass becomes thick enough, after years of accumulation, the ice flows outward under its own weight, like road tar on a hot day. As it moves, it carries with it rocks, soil, and other debris that it collects along the way. Glaciers grow wherever winter snow accumulation exceeds spring and summer losses, but the reason for net accumulation differs from place to place. In areas like the American Pacific Northwest, heavy precipitation rather than extreme cold causes glaciers to form at high elevations. In Antarctica low temperature is responsible; with only a few inches of snowfall each year, the continent qualifies as a desert.

Glaciers also die in different ways. Some, like the modern glaciers that fill valleys in coastal Alaska, flow down to the sea, where chunks of them, undermined by melting and wave action, break off to form icebergs—a process that glaciologists call, picturesquely, "calving." The immense continental ice sheets that covered much of the Northern Hemisphere between 2 million and about 10,000 years ago were stopped in their relentless southward march when they reached regions that were sufficiently warm that their flow was balanced by melting. These continental glaciers retreated—died—when the climate began to warm toward present conditions. They left behind the ridges and sheets of boulder-filled debris called moraines that now cover much of the landscape in northern North America and Europe. In cold, dry Antarctica, where the temperature climbs above freezing only rarely and locally, melting contributes little to the wearing away, or ablation, of glacier ice, but the wind contributes a great deal. Gales blowing off the ice cap fill the air with dancing particles of ice that slowly grind down the surface of the ice sheet to reveal whatever rocks it contains.

A glacier, then, is like a huge conveyor belt, picking up material in the zone of snow accumulation and along its path, then dumping it in the zone of ablation. If meteorites are included in the cargo manifest, they too will appear in the ablation zone. Unfortunately, most glacial deposits contain very many more terrestrial rocks than meteorites. Hence, some additional factors are required to explain the dramatic concentration of meteorites in parts of Antarctica. Takesi Nagata of Japan's Institute of Polar Studies has suggested, and other researchers have confirmed, that the requirements for a concentration of meteorites are a lack of native rocks along the glacier's path and the presence of a barrier in the zone of ablation. The absence of abundant terrestrial rocks prevents dilution of the meteorite component in the ice— the problem with most glacial moraines, including the one on which I live—and the barrier slows the ice and makes it well upward, exposing deep layers to ablation. This process has occurred at many places in Antarctica, gathering together meteorites that fell over a large region and during a long period of time and presenting them in a small area for easy collecting.

The sheer number of Antarctic meteorites is one of three reasons why they have attracted the interest of meteoriticists all over the world. The 7,000 specimens recovered to date, representing perhaps 700 separate falls, have increased the number of known meteorites by more than 25 percent. Though most of the new specimens are types that are already well represented in our collections, the harvest from Antarctica also includes new samples of several rare varieties and some new ones, including the first chunks of lunar rock that man has been able to collect without benefit of a spacecraft. The prospect of more such novelties is in itself a good reason why men and women face extreme hardships to return to Antarctica year after year. Another reason is that although all of the specimens from the ice cap are finds, they are finds with a difference. Though some are badly corroded—weathering proceeds in Antarctica, albeit slowly—many are fresher than finds from elsewhere. These meteorites are in fact fresh enough to be used for detailed chemical, and even biological, studies.

The third and most important reason for our great interest in Antarctic meteorites is that they add a time dimension to our meteorite collections. Most of the meteorites now in museums fell a few to a few hundred years ago, and thus they tell us only what kinds of meteoritic material, and in what proportions, have struck the Earth in the very recent past. In contrast, radioisotopic studies of Antarctic meteorites show that those collected to date fell during a time span of perhaps a million years. Samples from other sites may push the range of fall times back as far as 16 million years, the estimated age of the ice cap. Thus the Antarctic meteorites may tell us how the meteoroid population has changed with time and how the flux of cosmic rays has varied.

Naming Meteorites

Canyon Diablo, Allende, Krasnojarsk, Mirzapur. Leafing through the *Catalogue of Meteorites* is like browsing through a rack of brochures at a travel agency, for meteorites have been found in, and are named for, places all over the world. The *Catalogue* can also provide entertainment for those with a passion for associations. For example, there are many "presidential" meteorites (Washington County, Jefferson, Adams County, Monroe, Coolidge, Harding County, and Cleveland, to name just a few. Mount Vernon also squeezes in here, though the meteorite of that name fell in Kentucky rather than Virginia). For Shakespeareans, there are Hamlet and Elsinora, and for Anglophiles, Victoria West, Balfour Downs, Gladstone, and Chamberlin [*sic*]. Some meteorite names even provide unintended humor—for example, Bjurböle, whose Finnish name ("bee-yur-*burl*-uh"), is often corrupted by American scientists to something like "beer-belly"; and Bununu, a Nigerian achondrite whose name sounds like an expletive from a Tarzan movie.

Each new meteorite is named for a locality or a permanent geographic feature near its point of recovery. In cases where two or more distinct meteorites—a chondrite and an iron meteorite, for example—are found in the same area and no other name is available, we use letters to distinguish them—for example, Little River (a) and Little River (b). If two meteorites are found together and we are not sure whether they are or are not related, we use numbers (thus, Barratta no. 1 and Barratta no. 2). A committee of the Meteoritical Society oversees the application of these conventions and approves new names and changes for old ones.

Although this system works well for the rest of the world, it fails utterly for Antarctica, where there are many meteorites but few named landmarks and fewer towns. Fortunately, lunar scientists faced the same problem—too many samples from too few places—15 years ago and showed us how to handle it. Following their lead, we give each new Antarctic meteorite an abbreviated geographic name and a five-digit number. Thus the fifth sample curated (005) that was collected in the Allan Hills (ALH) by field party A in 1977 is called, simply, ALHA 77005. This scheme has shortcomings—the names are hard to remember, and lectures about Antarctic meteorites, like those about moon rocks, often sound like a play-by-play description of a Bingo game—but it is a rational solution for a real problem. For those who prefer the romance of the old system, there are almost 2,000 names in the latest *Catalogue,* and a few more appear in each new edition.

Meteorite Collections

Not only do meteorites fall everywhere, but they have found their way into collections that are scattered all over the world. There are major American collections in Washington (U.S. National Museum), New York (American Museum of Natural History), Chicago (Field Museum), and Tempe (Arizona State University) and important university collections at Harvard, Yale, UCLA, and the University of New Mexico. Major collections overseas include those in London (British Museum), Vienna (Museum of Natural History), Paris (National Museum of Natural History), Moscow (Academy of Sciences), and Calcutta (Geological Survey of India).

In our competitive society, one is tempted to try to arrange these collections in order of size or quality, but this would serve no useful purpose: some huge collections consist mainly of finds, and some small ones include a few unique and important meteorites. The point is that meteorites are scattered so widely that a meteoriticist must negotiate for samples with curators all over the world, a fact that gives meteorite research a decidedly international flavor.

There is also a measure of inefficiency in this system, and the risk that some curators will do a poorer job of preserving specimens than others. These problems have led to attempts to centralize our collections. For example, all meteorites that fall or are found in the United States are included under the National Antiquities Act and must be given or sold to the Smithsonian Institution in Washington. This law has led to some hard feelings and even occasional legal battles, but it has, in my opinion, done more good than harm. The Antarctic meteorites are still more centralized: they are housed in a modern facility in Houston, Texas, and a committee supervises their distribution to researchers. It is a good system in scientific terms. Those who find that dealing with a committee lacks the charm of negotiating a shopping list with a half-dozen curators can take comfort from the fact that some 2,000 meteorites—and all of the precious falls—remain scattered all over the world.

3

Types of Meteorites

Allende and Canyon Diablo illustrate that meteorites differ widely in composition and mineralogy as well as size. One important goal of meteorite research is to determine just how many kinds of meteorites there are; a second is to establish which kinds are related and how. To date, we have made more progress toward the first goal than we have toward the second: it is much easier to distinguish different kinds of meteorites than it is to demonstrate relationships among them.

An important reason for this difficulty is that meteoriticists cannot now visit the meteorite parent bodies. A geologist who is interested in relationships among the various volcanic rocks of Hawaii studies them first in the field and thus knows, long before he or she retreats to the laboratory, which rock types occur together and which do not, as well as which formed early and which formed late. These field observations help the geologist to identify *associations* of related rocks, which are much easier to understand than a random assortment of samples.

In the absence of field evidence, meteoriticists have devised clever ways to determine which meteorites formed together; we shall see examples of the resulting *meteorite associations* in Chapters 8 and 9. There is no doubt, however, that our work would be much easier if we could visit the meteorite parent bodies.

Stones, Irons, and Stony-Irons

By long-standing convention, we assign all meteorites to three broad divisions on the basis of their contents of two kinds of material:

24

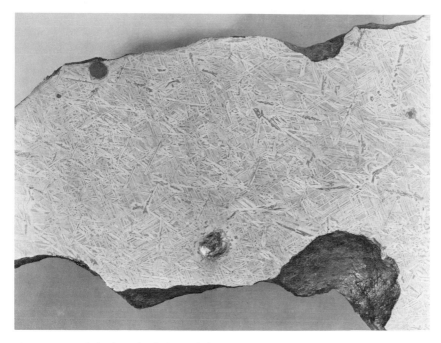

Figure 3.1 Polished, etched slice of the Mount Edith iron meteorite, showing its Widmannstätten structure and scattered nodules of iron sulfide (black). The field of view is 8 inches wide. (Photograph by Thane Bierwert, courtesy of the Department of Library Sciences, American Museum of Natural History; negative no. 34311.)

metallic nickel-iron (or "metal") and silicates, which are compounds of other chemical elements with silicon and oxygen. As their name suggests, the *iron meteorites,* for example Canyon Diablo, consist almost entirely of metal (Figure 3.1). At the opposite extreme, the *stony meteorites,* for example Allende, consist chiefly of silicates and contain little or no metal (Figure 3.2). A third category, *stony-irons,* includes those meteorites that contain similar amounts of metal and silicates (Figure 3.3). Since meteoritic metal weighs more than twice as much as the same volume of meteoritic silicates, we can usually distinguish these three kinds of meteorites by density, without more elaborate tests.

We can also subdivide the stony meteorites into two categories using nothing more complicated than a magnifying glass. The great majority of such meteorites, Allende included, are *chondrites* (pronounced "*kon*-drites"), which take their name from tiny, rounded objects—*chondrules*—that occur in most of them and are among their most puzzling features. The rest of the stony meteorites lack

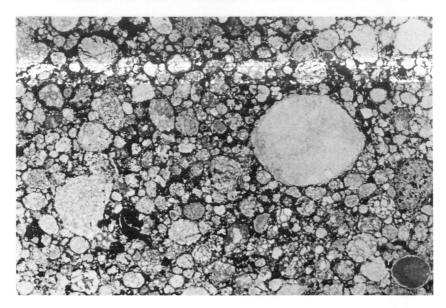

Figure 3.2 Photomicrograph of a thin slice of the Manych chondrite (ordinary, LL group) showing the characteristic aggregate texture of chondrites. The field of view is slightly more than 1/2 inch wide. (From Dodd, 1981.)

chondritic texture and are therefore called *achondrites*. The stony meteorite in Figure 3.2 is a chondrite. Since achondrites vary widely in texture, composition, and history, it is not feasible to illustrate them here, though some will be shown in later chapters.

As Table 3.1 shows, these four kinds of meteorites—irons, stony-irons, chondrites, and achondrites—are by no means equally abundant among observed falls: chondrites are much more common than all other kinds of meteorites put together. The reader who has seen meteorites in museums may find the numbers in this table surprising, for it shows that the irons, which are usually prominent in museum displays, are really quite uncommon. Curators like to highlight iron meteorites because many of them are large and their internal structure is spectacular in polished, etched slices (see Figure 3.1). Stony meteorites have a beauty of their own, but it only appears under the microscope: to the unaided eye, they appear to be—indeed they are—rather homely, gray or black rocks.

Thus a person who finds a meteorite in the field can assign it to one of four categories with little difficulty. To go further with meteorite classification, we have to be more specific about the minerals that make up a meteorite: *which* silicates are present, and *what kind* of metal? To answer these questions, we need to see more detail than is visible to the unaided human eye. Fortunately, thanks to Henry C. Sorby, an outstanding British mineralogist of the last century, we have the means to do so.

If a sample consists mainly of silicates, we make use of the fact that such minerals become transparent when sliced to about the thickness of a sheet of paper. We cut a flat face on the specimen, glue it to a glass slide, slice off most of it, and grind the rest to the desired thickness. Under a microscope, the resulting thin section looks like the one shown in the left-hand side of Figure 3.4. The minerals, most of them now transparent, have shapes, textures, and, in some cases, colors that enable us to identify them. To make identification easier, we can polarize the light that enters the specimen and polarize it again when it emerges. This sleight of hand gives some minerals characteristic colors, while others acquire telltale patterns of light and dark that are related to their structure.

A glance at the left-hand photograph in Figure 3.4 shows that thin sections have limitations, for some of the minerals in the specimen are

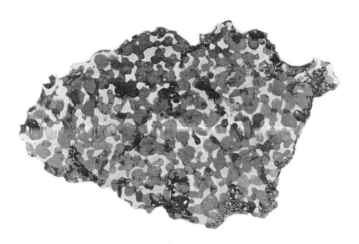

Figure 3.3 Polished, etched slice of the Thiel Mountains meteorite from Antarctica (see Figure 2.5). The specimen, an olivine stony-iron or pallasite, is about 6 inches wide and consists almost entirely of olivine (dark gray) and metallic nickel-iron. (Smithsonian Institution photograph, reproduced with permission.)

Table 3.1 Well-classified meteorite falls

Meteorite type	Number of falls	Percentage of total number
Stony		
Chondrites	602	87.4
Achondrites	57	8.3
Total	659	95.7
Iron	22	3.2
Stony-iron	8	1.2

Source: After Wasson (1974).

opaque even when sliced very thin. The metal and iron sulfide grains in this meteorite transmit no light and so must be studied in another way. Fortunately, Sorby encountered this problem too and found a solution: to view opaque minerals, one can polish their surfaces and observe them in reflected light. We can either polish a chunk or slice of a meteorite—the usual procedure for iron meteorites—or we can

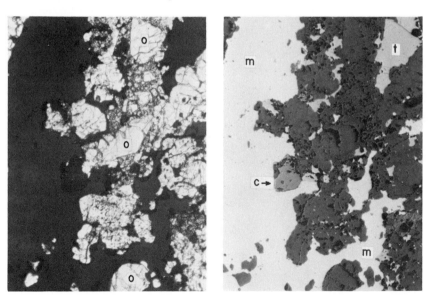

Figure 3.4 Polished thin section of the Antarctic chondrite ALHA 79039, photographed in transmitted light (*left*) and reflected light (*right*). The same area, 1 mm wide, is shown in both views. (*left*) Transparent parts of the section include several large olivine crystals (o); the balance consists of pyroxene and fine-grained intergrowths of pyroxenes and feldspar. (*right*) In reflected light, the silicates are unresolved and black. Opaque minerals vary in brightness and include nickel-iron (m, brightest), troilite (t, gray) and the iron-chromium oxide, chromite (c, dark gray).

Table 3.2 The most common meteoritic minerals

Mineral	Description	Composition
Nickel-iron	Alloys of iron, nickel, and cobalt. Occurs in two structures, kamacite and taenite. Present in most meteorites and dominant in irons.	Fe, Ni, Co
Troilite	Principal meteoritic sulfide, present in most meteorites.	FeS
Pyroxenes	Common iron-magnesium-calcium silicates. Many compositional subdivisions, e.g., enstatite, bronzite, hypersthene, diopside, augite. Present in most chondrites and achondrites and some stony-irons. Rare in irons.	$(Fe, Mg, Ca) SiO_3$
Olivine	Common iron-magnesium silicate, present in most chondrites, some stony-irons and achondrites. Rare in irons.	$(Fe, Mg)_2 SiO_4$
Plagioclase feldspar	A sodium-calcium-aluminum silicate. Present in most chondrites, many achondrites and some stony-irons. Rare in irons.	$NaAlSi_3 O_8$ to $CaAl_2 Si_2 O_8$

just polish the top surface of a thin section, making it possible to study both silicates and opaque minerals by switching back and forth between transmitted and reflected light. The right-hand side of Figure 3.4 shows the same area as in the left-hand photograph, but in reflected light.

Part of the basic documentation of each newly discovered meteorite is the preparation and description of thin or polished sections. With these, we can identify the minerals present and use their textures to learn something about the meteorite's history. The most common minerals found in meteorites are listed in Table 3.2.

When I was in school I spent many hours measuring the optical properties of minerals to estimate their chemical compositions: I could tell, for example, whether the iron-magnesium silicate olivine (perhaps more familiar to the reader as peridot, a semiprecious gem) was closer to the magnesian end of its compositional range or closer to the iron end. Although I finished school only two decades ago, that part of my education is no longer of much use to me, for this approach to mineral composition has been swept aside by a quite remarkable instrument called an electron probe microanalyzer or, less formally, a microprobe. The microprobe focuses a tiny beam of high-

Table 3.3 Mineralogical classification of meteorites

I. Stony meteorites
   A. Chondrites
      1. Carbonaceous (four subdivisions or groups)
      2. Ordinary (three groups)
      3. Enstatite (two groups?)
   B. Achondrites
      1. Calcium-poor
         a. Enstatite (aubrites)
         b. Hypersthene (diogenites)
         c. Olivine (chassignite)
         d. Olivine-pigeonite (ureilites)
      2. Calcium-rich
         a. Augite (angrite)
         b. Diopside-olivine (nakhlites)
         c. Pyroxene-plagioclase ("basaltic")
            (1) Eucrites
            (2) Howardites
            (3) Sherghottites

II. Stony-iron meteorites
   A. Olivine (pallasites)
   B. Pyroxene-plagioclase (mesosiderites)
   C. Bronzite-tridymite (siderophyre)
   D. Bronzite-olivine (lodranites)

III. Iron meteorites
   A. Hexahedrites
   B. Octahedrites
   C. Ataxites

energy electrons on the surface of a sample and uses the x-rays that it produces to tell which chemical elements are present and in what proportions. Because we can focus the electron beam to a spot about one-thousandth of a millimeter in diameter (barely a thousandth the size of the period at the end of this sentence), we can analyze individual grains that are quite invisible to the naked eye. Moreover, we can see what we have analyzed, since the microprobe is equipped with a microscope whose cross-hairs mark the location of the beam. Though the meteoriticist's arsenal now includes still more sophisticated tools, it is safe to say that the microprobe has done more than any other instrument to pry out the secrets of meteorite mineralogy.

On the basis of microscopic study, we can expand the simple four-fold classification of meteorites to the much more elaborate one given in Table 3.3. Most of the subdivisions of stony and stony-iron meteorites in this table have names that identify the most important mineral or minerals present. Since the pyroxenes are important in many meteorites, it is not surprising that many of the mineralogical names (enstatite, hypersthene, pigeonite, augite, diopside, bronzite) refer to different species of pyroxene. The iron meteorites, composed almost entirely of nickel-iron, are named according to the metal's texture, which reflects its nickel content and history.

Some groups of meteorites also have names that honor a prominent meteoriticist (for example, the howardites, named for Edward Howard) or the first meteorite of its kind that was discovered (for example, the nakhlites, named for the Egyptian meteorite Nakhla). Indeed, some groups have more than one common name; paradoxically, this is most often true of rare types. Table 3.3, forbidding as it is, lists only those names that are in common use.

To summarize, then, we can divide meteorites into four categories on the basis of density and simple observations with nothing fancier than a pocket magnifier. With thin or polished sections and a microscope, we can recognize many important subdivisions. Since these tools have been available for more than a century, it is not surprising that Table 3.3 has changed little since Sorby's day.

What has changed, and dramatically, is our understanding of what the differences among various kinds of meteorites mean. For one thing, we now know that some of the groups of meteorites in Table 3.3, though very different in chemical composition and mineralogy, formed together. For example, three kinds of achondrites (eucrites, howardites, and diogenites) and one kind of stony-iron meteorite (mesosiderites) are very closely related and apparently come from the same parent body. Most pallasites and one kind of iron meteorite are likewise related, and so, apparently, are the shergottites, nakhlites, and chassignites. These relationships will be discussed in Chapters 8 and 9. We have also come to realize that of all the boundaries in Table 3.3, the most important is that between chondrites and all other meteorites. Not only are chondrites the most common of meteorites, but they differ from other meteorites—and indeed all other rocks—in other ways that make the chondrites uniquely important to the story of the solar system.

Sun Samples

Television commercials for a popular instant camera claim that it comes equipped with "a piece of the sun"—a built-in electronic flash gun that helps the camera compensate for poor light. In one such commercial, an actor reaches up and snatches a brilliant bit of light, to his partner's amazement. It is an effective commercial, but inaccurate. The actor should reach *down*, not up, and he should pick up a small, gray rock: a chondrite. For chondritic meteorites are the closest things we have to pieces of the sun, and some of them are very close indeed.

This is quite an astounding statement, and it will probably strike many readers as obviously wrong: as every schoolchild knows, the sun consists of hot gases, the most abundant of which, by far, is hydrogen. Clearly chondrites do not consist of hydrogen. In fact they contain very little of this element and other elements—helium, nitrogen, and argon, for example—that are gases under most conditions of temperature and pressure. Just what do we mean when we say that chondrites chemically resemble the sun?

If we consider the abundances of most elements, gases excluded, relative to a suitable reference element, for example silicon, we find that the element/element ratios we measure for chondrites are almost identical to the ratios astronomers have measured for the sun. The

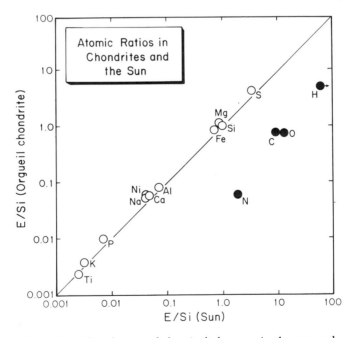

Figure 3.5 Abundances of chemical elements in the sun and in carbonaceous (CI) chondrites, relative to silicon. If the solar and chondritic element/Si ratios were identical, all points would fall on the 45-degree line shown in the figure.

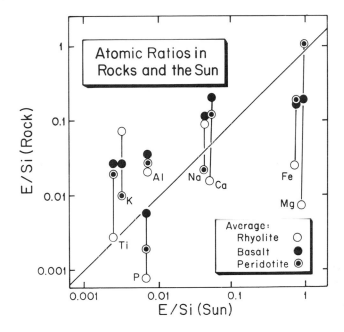

Figure 3.6 Abundances of chemical elements in the sun and in three common terrestrial rocks, relative to silicon.

easiest way to show this is by plotting element/silicon ratios for chondrites against the same ratios for the sun. As Figure 3.5 shows, the points for most of the elements shown fall close to a 45-degree line— the line they would describe if each ratio were exactly the same in chondrites and the sun. The only exceptions are elements—for example, nitrogen (N), carbon (C), oxygen (O), and hydrogen (H) that are usually gases, either by themselves or when combined with other elements. We could add many more elements to Figure 3.5 without changing the result: as long as we restrict the comparison to elements that remain solid to temperatures of a few hundred degrees Celsius, chondrites in general and one variety in particular have almost solar compositions.

   This relationship is remarkable, the more so because it is unique to chondrites. Figure 3.6 makes the same comparison for three important terrestrial rocks: rhyolite, a volcanic rock chemically equivalent to the granite that forms the cores of mountain ranges and, in smaller pieces, decorates public buildings and cemeteries; basalt, a volcanic rock that makes up most oceanic islands and covers the ocean floors; and peridotite, an olivine-pyroxene rock that commonly appears as chunks in basaltic lavas and is probably similar in composition to the Earth's second layer or mantle. The figure shows that these three rocks vary a great deal among themselves, but all lie far from the solar composition. All rocks, nonchondritic meteorites included, differ from the chondrites in that various processes—melting and crystallization foremost—have worked to drive their chemical compositions away from that of the sun. For this reason, we call nonchondritic materials *differentiated* to set them apart from the chemically *primitive* chondrites.

This basic distinction between chondrites and other meteorites brings us to a fork in the road. Differentiated meteorites, like the rocks that we find on the Earth and the moon, carry invaluable records of the way in which the planets evolved, but they tell us next to nothing about how planets formed. When a body melts and differentiates, it effectively burns its birth certificate and shreds its baby pictures. Chondrites, alone among the rocks that we can collect and study, carry a record of the first steps in the history of the solar system.

The next three chapters will focus on the chondrites—what they are, where they come from, and where they formed. In later chapters we shall double back to consider what differentiated meteorites add to the story.

4

Three Perspectives
on Chondrites

It takes many people with different skills to work out the history of meteorites. Those of us who were trained in geology study meteorites in much the same way that we study rocks from the Earth or the moon: we slice or polish them, examine them with a microscope, and analyze their minerals with a microprobe. Others perform chemical analyses, either by dissolving meteorites, treating the solution with reagents, and weighing the solids thus formed—a smelly pursuit familiar from high school chemistry—or by applying various instrumental techniques. Still others go beyond tallying the elements present in a meteorite to examine their isotopes: atoms of the same element that have different weights. The challenge of meteorite research, and one of its pleasures, is juggling data from these and many other fields of science—astronomy, metallurgy, even biology—to piece together a story that makes sense.

This chapter will look at the chondrites from the viewpoints of geology, analytical chemistry, and nuclear chemistry, using the testimony of these three disciplines to form a picture of these primitive meteorites. Since chondrites are rocks, albeit odd ones, we will look at them first through the eyes of the geologist. What kind of rocks are they?

Through the Microscope

At some point in your schooling you probably learned that all rocks are either *igneous, sedimentary,* or *metamorphic.* Igneous rocks form when molten rock or *magma* cools and crystallizes. If crystallization takes place at the surface of the Earth or any other body, the rock that results is called *volcanic;* if it takes place below the surface, the resulting rock is called *plutonic.* Because magma crystallizes very rapidly at the Earth's surface, the minerals within volcanic rocks have little time to grow and most such rocks are very fine-grained; a good example is the black basalt that makes up much of Hawaii and many other oceanic islands. Slower cooling and crystallization endow plutonic rocks with larger crystals and give them a coarser texture. Granite is a familiar plutonic igneous rock.

Thus the common denominator of igneous rocks is that they form at very high temperatures and from molten material. At the opposite extreme, sedimentary rocks form in the cool environment of the Earth's surface, either by settling of mineral grains from water or air or by crystallization of minerals from a solution, for example seawater. Sandstone and shale are common particulate sedimentary rocks, the former made from sand and the latter from mud. Rock salt is a good example of a sedimentary rock that forms from a solution. Some limestone also forms in this way, though much is produced with the help of corals, clams, and other shell-forming organisms.

When an igneous or sedimentary rock is subjected to high temperature, high pressure, or both, its texture and mineralogy change to suit its new environment. Thus a fine-grained, sedimentary limestone may become a coarse-grained marble, a shale may become a slate, and a granite may become a banded gneiss. The processes involved in these changes are called metamorphism, and the products are called metamorphic rocks. Though metamorphism includes many processes and produces varied products, it usually blurs the original texture of a rock and makes it hard to tell just what it was like.

Some terrestrial rocks fit comfortably in one of these three pigeonholes, but others sprawl across two or three of them. Thus the ash thrown from a volcano—Mount St. Helens, for example—may be carried by the wind and deposited to produce a form of sediment. Geologists call such a deposit pyroclastic to acknowledge that it is a sedimentary rock composed of igneous material.

Chondritic meteorites spread across all three of the categories, as we can see by looking at a few of them under a microscope. The meteorite shown in Figure 4.1, like most other chondrites, consists of tiny spheres and fragments of silicate material (*chondrules*), bits of metal and iron sulfide (opaque and therefore black in the photograph), and fine-grained material (or matrix) that lies between these materials. Obviously this chondrite is a sedimentary rock of the particulate type, for we can see that it consists of innumerable individual particles bound together by a matrix. In fact, if we look only at its texture and ignore its mineralogy, the meteorite in Figure 4.1 looks remarkably like many terrestrial sandstones.

To call the chondrite in Figure 4.1 a sedimentary rock is accurate but incomplete, for it also had an igneous history. If we examine individual chondrules within it, we find that their minerals (olivine, pyroxenes, glass) and their textures (globules with delicate fans of pyroxene crystals; droplets and fragments with stumpy olivine and pyroxene crystals set in glass) are like those that we find in rapidly cooled igneous rocks. We conclude, then, that this chondrite is a sedimentary rock made of many bits of igneous rock.

But we are still one step short of the whole truth. Though nothing in the photograph advertises it, this meteorite also experienced metamorphism. If we look carefully at the fine-grained matrix between the chondrules, using a light or electron microscope, we find that its crystals have begun to grow together; when we analyze these matrix crystals with the electron microprobe, we find that each mineral is approaching one composition, as minerals do during metamorphism. This chondrite is, then, a lightly *metamorphosed sedimentary* rock whose component particles were formed by *igneous* processes. Using nothing more than a microscope and some common sense, we have already pieced together a rather complex history for this meteorite.

Metamorphism has touched the chondrite in Figure 4.1 so lightly that its original texture is almost perfectly preserved. In contrast, the consequences of metamorphism are obvious in Figure 4.2. In this chondrite, individual chondrules are barely perceptible because their crystals have grown together with those of the matrix. If we analyze the olivine and pyroxenes in this meteorite, we find that they are homogeneous and have chemical compositions that indicate metamorphic temperatures of several hundred degrees Celsius. Some chondrites have gone still further along the metamorphic path and lack chondritic structure altogether. In fact, a few such meteorites were at first classified as achondrites until chemical analyses showed where they really belong.

These two examples show that many chondrites have igneous, sedimentary, *and* metamorphic histories. The chondrules and metal and troilite grains within them formed from molten material; these particles were brought together by some kind of sedimentary process; and the resulting sedimentary rock was metamorphosed at high temperatures.

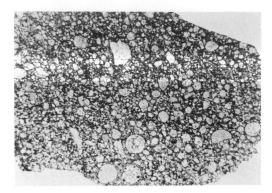

Figure 4.1 Thin section of the Selma, Alabama, meteorite, a weakly metamorphosed ordinary chondrite (H-group) in which chondrules are sharply defined against a dark, fine-grained matrix. The section, viewed in transmitted light, is 0.7 inches wide. (From Dodd, 1981.)

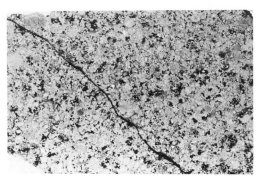

Figure 4.2 Thin section of Colby, a strongly metamorphosed ordinary chondrite (L-group) in which chondrules are indistinct in a coarse-grained, clear matrix. A vein of shock-melted material crosses the thin section, which is viewed in transmitted light and is one inch wide. (From Dodd, 1981.)

Some chondrites, one of which is shown in Figure 4.3, add a fourth chapter to this story, for after they were formed and metamorphosed, they were shattered and jumbled together with other materials to form a fragmental rock called *breccia* (pronounced "*bretch*-ee-uh"). Some chondritic breccias consist of fragments of a single chondrite, but many throw together different kinds of material. Indeed, a breccia can be a miniature rock collection: when I studied the brecciated St. Mesmin chondrite a few years ago, I looked at more than a dozen thin sections before I found two that looked alike.

What process produced meteoritic breccias? The minerals within them bear fractures and other structures that we also find in rocks around terrestrial and lunar meteorite craters. Thus we conclude that meteoritic breccias formed when meteoroids struck their parent bodies at very high speeds. As we shall see, such impacts played a very important role in the evolution of meteorites.

It is obvious from the examples in Figures 4.1 to 4.3 that chondrites differ a great deal in appearance and have quite varied histories. Since metamorphism and impact shock tend to rub out evidence for a chondrite's earlier history, trying to use a meteorite like the ones in

Figures 4.2 and 4.3 to study chondrules is like trying to learn about flour and eggs by studying a cake. Fortunately, a few chondrites, like the one in Figure 4.1, escaped such extensive processing. Such meteorites are uniquely valuable, for they remember stages of their history that other chondrites have forgotten.

There is one more important variation on the chondritic theme. Though chondrites take their name from chondrules, a handful of these meteorites, one of which is shown in Figure 4.4, are chondrule-free. Chemical analyses of such meteorites show that they are indeed chondrites and in fact chemically resemble the sun more closely than the other meteorites that have been discussed. It is one of the many small ironies of meteoritics that chondrules—the hallmark of chondrites—are absent from the most chondritic meteorites of all.

In summary, then, chondrites are the products of a history that included—reading backward in time—shock metamorphism and brecciation; thermal metamorphism; accretion of particles; and for-

Figure 4.3 Slice of the Bloomington, Illinois, meteorite, a brecciated ordinary chondrite (LL-group). The meteorite is 1.3 inches wide. (From Dodd, Olsen, and Clarke, 1985, reproduced by permission of *Meteoritics*.)

Figure 4.4 Thin section of the Orgueil carbonaceous chondrite (group CI), which lacks chondrules and consists principally of very fine-grained hydrous silicates and iron oxide (magnetite). Veins of carbonates and sulfates (*white*) were deposited from water in the meteorite's parent body. The scale bar, calibrated in micrometers, is 0.008 inches long, and the meteorite is viewed in transmitted light. (From Richardson, 1978, reproduced by permission of the author and *Meteoritics*.)

mation of those particles. Some chondrites passed through all of these stages, others just the early ones. Since it is clear that many chondrites are very experienced rocks, why do we call them "primitive"? We do so because the chondrites, almost without exception, escaped the one process—melting—that would have drastically changed their chemical compositions, destroyed their original textures, and erased all evidence of their early history. Though many chondrites had lively physical histories, their chemical compositions remain much as they were when they formed: they are *physically* evolved but *chemically* primitive rocks that preserve a chemical and isotopic record of events in the early solar nebula. As we have seen, a precious few chondrites are also physically primitive, preserving textural and mineralogical records as well.

We can learn a lot about the history of chondrites by looking at them under a microscope, but some important questions remain unanswered. For example, the meteorites in Figures 4.1 to 4.4 look quite different. Are the differences solely a result of varied physical histories, or are there basically different kinds of chondrites? To answer this question, we have to leave the realm of geology and turn to that of analytical chemistry.

Chemical Groups of Chondrites

When a 56-pound chondrite fell at Wold Cottage in Yorkshire, England, in 1795, Sir Joseph Banks, who was then president of Great Britain's Royal Society, noted that the stone looked very much like two other recently fallen meteorites. To find out whether the similarity went beyond appearance, Banks asked a young chemist, Edward Howard, to analyze all three stones. Howard's analyses, prepared in collaboration with an exiled French mineralogist, Jacques-Louis Comte de Bournon, and published in 1802, confirmed Banks's suspicion that the three chondrites were chemically similar.* Analyses by Vauquelin in Paris and Klaproth in Berlin soon extended this conclusion to several other chondrites.

As meteorite analyses accumulated during the nineteenth century, it became evident that "similar" did not mean "identical." Chemists and mineralogists noted that despite their general chemical similarity, chondrites differ widely in the way in which their iron is distributed. In some, nearly all of this element occurs in metal and sulfide, and the accompanying silicates are nearly iron-free. Because enstatite, a magnesium pyroxene, is the most abundant mineral in such meteorites, they are called *enstatite chondrites*. At the opposite pole from the enstatite chondrites are meteorites, for example the one shown in Figure 4.4, that contain little or no metal and in which iron occurs in sulfides, silicates, and magnetite, the magnetic iron oxide. Because most such meteorites also contain abundant carbon, we call them *carbonaceous chondrites*. A third class of meteorites lies between the enstatite and carbonaceous chondrites. These, which include the meteorites shown in Figures 4.1 to 4.3, contain both metallic iron and iron-bearing silicates. Since most chondrites fall into this category, we call them *ordinary* or—in Great Britain—*common* chondrites.

It is clear from their different iron distributions that these three classes of meteorites formed in different chemical environments. The enstatite chondrites formed under such oxygen-poor conditions that they contain minerals that we never see on Earth, for example, oldhamite, the calcium sulfide, and niningerite, the magnesium sulfide. In contrast, the carbonaceous chondrites formed in an oxidizing environment and the ordinary chondrites formed somewhere between these extremes. Until quite recently it appeared that chemical differences among these three kinds of chondrites are limited to oxygen, water, and other constituents that vaporize at low temperatures— that all chondrites consist of the same basic material exposed to different degrees of oxidation, like a new iron pipe and its old, rusty

---

*A paper written by Howard and de Bournon in 1802 played a large role in persuading the scientific community that meteorites come from beyond the earth. Howard's name lives on in the *howardites;* de Bournon's is familiar to mineralogists from the lead-copper-antimony mineral *bournonite.*

equivalent. Indeed, as the great British meteoriticist G. T. Prior noted in 1916, variations of mineral composition and metal abundance from class to class suggest such a relationship: where metallic nickel-iron is abundant, it is iron-rich and the associated silicates are iron-poor; where there is little metal, it is nickel-rich, and much of the iron in the meteorite resides in the silicates, olivine and pyroxene. It appeared to Prior that one might convert one kind of chondrite to another just by redistributing iron among silicates and metal.

As we all must, Prior made the best explanation he could with the chemical data at his disposal. However, we now know that chemical differences among chondrites go beyond oxygen and other easily vaporized constituents to include elements that vaporize only at very high temperatures. This first became evident in 1953, when Harold Urey and Harmon Craig published the results of a careful review of the meteorite analyses that had appeared to that time. After they weeded out all faulty analyses, they found that the iron contents of chondrites tend to fall near either 22 percent or 28 percent by weight (Figure 4.5). Urey and Craig concluded from this that there are two

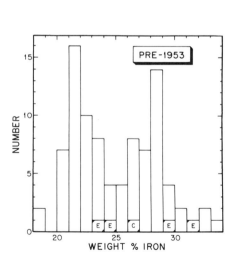

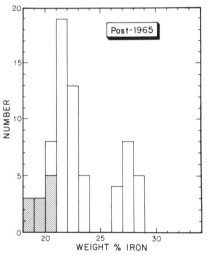

Figure 4.5 Distribution of iron abundances in chondritic meteorites: analyses obtained prior to 1953 and judged to be superior by Urey and Craig (1953). Several enstatite chondrites (E) and one carbonaceous chondrite (C) are marked. The other analyses refer to ordinary chondrites.

Figure 4.6 Distribution of iron abundances in chondritic meteorites: analyses of ordinary chondrites prepared by Eugene Jarosewich between 1965 and 1984. The hatched "tail" on the low-iron peak in the histogram refers to chondrites of the LL-group.

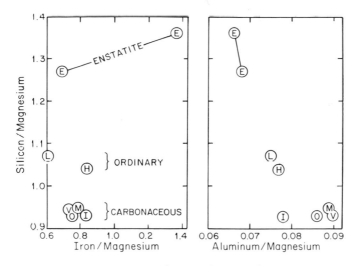

Figure 4.7 Atomic ratios of iron, silicon, and aluminum to magnesium in chondrites of various classes and groups. Averages are shown for all but the enstatite chondrites, which vary between the limits shown in the figure. Ratios for I group carbonaceous chondrites are very close to those observed for the sun. (Compositions are from Dodd, 1981.)

distinct kinds of chondrites: a high-iron (or "H") group, which includes the carbonaceous and some ordinary chondrites; and a low-iron ("L") group, which includes the rest of the ordinary chondrites. Curiously, the enstatite chondrites are distributed more or less equally between these iron groups.

As more chemical analyses appeared, the differences in iron con-tent that Urey and Craig noted became sharper (Figure 4.6) and other differences became evident as well. Careful analyses by Louis Ahrens and his co-workers at the University of Capetown showed, for example, that the carbonaceous, enstatite, and ordinary chondrites have quite different silicon/magnesium ratios.

If we compare the abundances of iron, magnesium, and silicon—after oxygen, the three most abundant elements in chondrites—we find that chondrites fall naturally into three large classes and eight smaller groups (Figure 4.7). Four groups (I, M, O, and V) that cluster together near the sun's composition in Figure 4.7 make up the carbonaceous chondrites. Three more groups (H, L, LL), with higher silicon-magnesium ratios, constitute the ordinary chondrites. Each of the seven groups of ordinary and carbonaceous chondrites is quite homogeneous and distinct. In contrast, members of the eighth (E) group have compositions that are scattered between the extremes shown in Figure 4.7. This wide chemical variation, peculiar to the enstatite chondrites, has led some meteoriticists to suggest that they really constitute two groups. Others, including myself, believe that all enstatite chondrites formed together but under unusual circumstances.

It is clear from Figure 4.7 that chondritic material is less homogeneous than earlier workers supposed. How do the chemical differences shown in that figure match up with the physical variations shown in Figures 4.1 to 4.4? Shock features are scattered more or less randomly among members of all chondrite groups, but metamorphic effects are distributed quite systematically (Figure 4.8). Most of the carbonaceous chondrites show evidence of weak metamorphism or none. On the other hand, each group of ordinary chondrites spans a wide range from unaltered to severely metamorphosed material. The enstatite chondrites may cover a still wider range: their chemical compositions, textures, and mineralogy suggest that some reached temperatures that were high enough to melt them, wholly or in part. I suspect that such melting is responsible for the chemical differences among enstatite chondrites, though most meteoriticists now prefer the view that these meteorites constitute two groups.

In addition to indicating that the various groups of chondrites have different physical histories, Figure 4.8 carries another important message: since each group includes at least a few meteorites that did not undergo metamorphism, it is clear that the chemical differences that

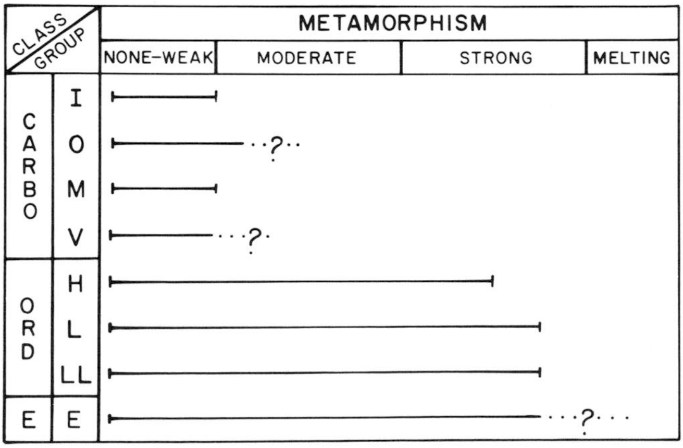

CLASS / GROUP		METAMORPHISM			
		NONE–WEAK	MODERATE	STRONG	MELTING
C A R B O	I	⊢————⊣			
	O	⊢————————··?··			
	M	⊢————⊣			
	V	⊢————···?·			
O R D	H	⊢————————————————⊣			
	L	⊢——————————————————⊣			
	LL	⊢——————————————————⊣			
E	E	⊢——————————————————————···?···			

Figure 4.8 Variations of metamorphic intensity in chondrites of various chemical classes and groups.

distinguish the groups arose before that process took place. Evidently the eight (or nine) kinds of chondrites have been distinct since their component materials settled out of the primitive solar nebula.

Thus analytical chemistry adds a great deal to our emerging picture of chondrites. It shows that even though all chondrites chemically resemble the sun and are in that sense primitive rocks, there are at least eight kinds of chondritic material with distinctive compositions. It shows further that these chemical differences arose very early and testify to processes in the primitive solar nebula. Clearly there are, and always have been, many kinds of chondrites.

The Antiquity of Chondrites

Even analytical chemistry leaves some questions dangling. For example, it tells nothing about *when* chondrites formed and where. To add dates to this history of the chondrites, we must turn to nuclear chemistry, the branch of science that deals with isotopes. It is clear from what we have seen thus far that even chondrites, the least experienced rocks known to us, have had a rather complex history. Reading backward from today, that history includes the following stages:

10. Recovery
 9. Terrestrial residence and weathering
 8. Transit in space
 7. Ejection from the parent body
 6. Impact reworking on the parent body
 5. Thermal metamorphism
 4. Accretion of the parent body
 3. Formation of chondrules, matrix, and so on
 2. Formation of the solar nebula
 1. Formation of the chemical elements

Some chondrites went through all of these stages, but others skipped some. For example, the carbonaceous chondrite shown in Figure 4.4 lacks chondrules (step 3), and this meteorite and many other carbonaceous chondrites experienced little or no metamorphism (step 5). On the other hand, some ordinary chondrites bear evidence of still more elaborate histories, with repeated episodes of impact reworking, burial, and thermal metamorphism that make them extremely hard to study and even harder to understand.

Meteoriticists would like to document every stage in the history of chondrites, but we are particularly interested in stages 1 to 5, which bear on the early history of the solar system and the origin of planets.

When did chondrules form? When did they and other materials come together to form parent bodies, and what were those bodies like? Isotopic chemistry answers some of these questions and gives partial answers to others.

Isotopes. Isotopes are atoms of a chemical element that contain different numbers of neutrons and hence have different atomic weights. Hydrogen (H) is the simplest example: though 99.99 percent of all hydrogen atoms consist of just one proton and one electron, a very small proportion of them also contain one neutron and a still smaller proportion contain two neutrons. Since each proton or neutron has a weight of approximately 1 atomic mass unit, the three hydrogen isotopes have atomic weights of 1, 2, and 3 units. Likewise, the most abundant oxygen isotopes, all with eight protons, contain eight, nine, or ten neutrons and have atomic weights of roughly 16, 17, and 18 units.

Many isotopes, for example hydrogen-1 and hydrogen-2 (deuterium) and the three common oxygen isotopes, are *stable:* once formed, they persist indefinitely. Other isotopes, for example hydrogen-3 (or tritium) are *unstable* or *radioactive:* they change or *decay* spontaneously to form other isotopes of the same or another element at a fixed rate. Since the extent of decay of a parent isotope to its daughter can be used to measure the passage of time, we use many unstable isotopes as radioactive clocks to date events in the histories of the Earth and other bodies in the solar system.

Figure 4.9 shows an important difference between radioactive clocks and other timepieces. In an hourglass, for example, sand passes from one globe to the other at a constant rate. Thus if half of the sand passes through in 5 minutes, all of it will pass in 10 minutes. In contrast, if half of the atoms of a radioactive isotope decay in one unit of time, half of the *remaining* atoms (or one-quarter of the original number) will decay during the next unit; half of those left after the second unit will decay during the third, and so on. The time it takes for half of the radioactive atoms to decay is called an isotope's *half-life*. Parents, daughters, and half-lives for several isotopic systems that we use for dating rocks and meteorites are listed in Table 4.1.

Complications of dating. Reading a radioactive clock is, in principle, very simple: we measure the abundances of a parent isotope and its daughter in a rock or mineral, using a sensitive instrument called a mass spectrometer; we then use the parent's half-life to calculate how much time has passed since the radioactive clock started ticking. However, many factors conspire to make radiometric dating much harder than it sounds and almost as much an art as a science. One

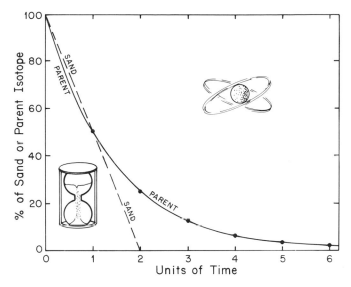

Figure 4.9 Comparison of the exponential decay of a radioactive isotope and the linear decay of an hourglass.

factor is that not all isotopic systems are suitable for all rocks. Clearly, a sample to be dated must contain a useful parent isotope, but beyond that, it must contain both parent and daughter in measurable amounts. This requires, rather paradoxically, that we know something about the age of a rock before we set out to determine it.

Two examples will show why this is so. Suppose we want to date an old terrestrial rock, for example a granite from the Adirondack Mountains, whose age is likely to be about a billion years. The fast clocks — carbon (C) 14, aluminum (Al) 26, and iodine (I) 129 — are useless for this purpose, for although these isotopes were probably present in the primitive Earth, they decay so fast that they had already decayed completely (that is, they were extinct) when our granite formed. To date this rock, we must turn to the other systems in Table 4.1, all of which decay fast enough to produce measurable amounts of their daughter isotopes in a billion years while leaving measurable amounts of their parents.

Suppose, on the other hand, that we want to date a lava flow that overran a Greek city 3,000 years ago. Here the long-lived systems fail us, because they have produced immeasurably small amounts of daughter isotopes during this very short period. Curiously, one of the fast clocks proves to be best in this case, for although the Earth's original complement of C-14 is long gone, this isotope is produced continuously in the upper atmosphere, circulates, and is taken up by living tissue. When an organism dies it stops accepting C-14, and the C-14 atoms within it start to decay. We can use the extent of their decay to tell when the organism—a tree, perhaps, or a Greek citizen—succumbed to the lava flow. Though radiocarbon dating reaches no farther back than about 50,000 years, the rapid decay of C-14 makes it very sensitive to small time differences between very young samples. Hence it is a mainstay for students of recent geological and human history.

Even after we choose an appropriate isotopic system for the rock that we wish to date, we may find it hard to tell just what the measured age means. If a metamorphic rock formed from a sedimentary parent that was itself made of igneous particles, does the age that we determine refer to metamorphism, sedimentation, or the igneous event that produced the particles? Some isotopic systems, especially those whose daughters are gases (potassium-argon for example), are so easily disturbed by heating that they will record only the most recent high-temperature event, in this case metamorphism. To see through that event to earlier ones, we must use clocks that are less easily reset. Ideally, we should use several isotopic systems to learn as much about the rock's history as we can.

Chondrite ages. Since 1955, when Claire Patterson first used lead isotopes to date the formation of chondrites and the Earth, we have known that both formed about 4½ billion years ago. Subsequent work with lead and other isotopic systems has refined that figure for chondrites to 4.55 billion years and has extended it to many other

Table 4.1 Major isotopic systems used as radioactive clocks

Parent isotope	Daughter	Half-life (years)
Carbon (C)-14	Nitrogen (N)-14	5,730
Aluminum (Al)-26	Magnesium (Mg)-26	740 thousand
Iodine (I)-129	Xenon (Xe)-129	17 million
Uranium (U)-235	Lead (Pb)-207*	704 million
Potassium (K)-40	Argon (Ar)-40	1.3 billion
Uranium (U)-238	Lead (Pb)-206*	4.5 billion
Thorium (Th)-232	Lead (Pb)-208*	14 billion
Rubidium (Rb)-87	Strontium (Sr)-87	49 billion

Note: Parent isotopes in the top part of the table are extinct: the entire quantity that was present in the Earth when it formed, 4.55 billion years ago, has decayed. Decay of the uranium and thorium isotopes (*) yields helium (He) as well as lead.

kinds of meteorites, though by no means all. A great deal of work has gone into trying to determine just what the 4.55-billion-year age of chondrites means—that is, whether it dates chondrule formation, accretion of the parent body, metamorphism, or all of these events. The results vary somewhat from one kind of chondrite to another. It appears that the evolution of ordinary and perhaps enstatite chondrites, from chondrule formation through accretion and metamorphism, took no more than 100 million years and perhaps much less. Chondrule formation itself probably took place during the first few million years of this interval, during or very shortly after the birth of the sun. The carbonaceous chondrites tell a slightly different story, for although the chondrules and other high-temperature materials within them formed very early, some isotopic systems testify that these meteorites continued to react with their surroundings for perhaps half a billion years—much longer than the entire history of the ordinary chondrites.

It is clear, then, that chondrites formed very early in the history of the solar system and reached their present form at least 4 billion years ago. Though some meteorites of each chondrite group yield much lower ages, all such objects bear evidence of strong impact shock, which affected many chondrites, on and off, long after their parent bodies ceased to be internally active.

Chondrites are so old and their active lives were so brief that it is hard to work out the details of those lives. The long-lived isotopes that we use to date chondrites decay so slowly that the uncertainty in an age of 4.55 billion years is almost as large as the time between the formation and metamorphism of the ordinary chondrites. Fortunately, chondrites formed early enough in the history of the solar system to incorporate small amounts of such short-lived radio isotopes as aluminum-26, iodine-129, and plutonium-244. Since these isotopes were decaying rapidly while the chondrites evolved, small time differences are reflected by large changes in the parent/daughter ratios. The iodine-129 and plutonium-244 clocks are rapidly filling in details of the brief histories of chondrites and other meteorites. I shall return to the early histories of chondrites in Chapter 10, where I discuss the use of meteorites in our efforts to understand the early solar system.

5

Sources of
Meteorites

Thomas Edison's definition of genius—90 percent perspiration, 10 percent inspiration—applies just as well to the collective genius of a science as it does to that of individual practitioners. Most research is hard, tedious work that adds paper to library shelves without producing much enlightenment. Only rarely does a field of science experience a short burst of insight that transforms mountains of unruly data into a bold new view of reality that deserves to be called a revolution.

Biology went through such a revolution more than a century ago, when Darwin and Wallace proposed the grand concept of organic evolution. Geology experienced one in the early 1960s with the discovery that huge plates of the Earth's crust slide over its mantle, pushed along by sea-floor spreading. Though neither the concept of evolution nor the plate theory still stands in its original form, both of these breakthroughs drastically changed the way we look at nature. They opened new doors to research even as they slammed old ones.

The science of meteoritics has passed through two revolutions, both of which involved dramatic shifts of opinion on where meteorites come from and what they mean. The first took place shortly after 1800, when British and French scientists demonstrated that iron and stony meteorites are related and that both kinds come from beyond the Earth. Because the realization that meteorites are samples from space led men to collect and study these objects in earnest, we mark this event as the birth of the modern science of meteoritics. The second revolution began about 25 years ago and continues today. Until the mid-1960s, most students of meteorites interpreted them as bits of the asteroids, small bodies—none as big across as Pennsylvania—that circle the sun between the orbits of Mars and Jupiter. They viewed the asteroids, in turn, as the remains of an Earth-like planet that once lay between Mars and Jupiter and was shattered in a cosmic catastrophe.

Chemical, isotopic, and astronomical data gathered during the last quarter-century have led us to accept the first part of the old interpretation but reject the second. We are now virtually certain that most meteorites come to the Earth from asteroids and that the few exceptions come from other bodies in the inner solar system. We are just as sure that the asteroids are *not* fragments of one large parent body, but rather the remains of a swarm of small objects that—for whatever reason—did not coalesce to form a tenth planet. As such, they give us insights into the birth and babyhood of the solar system that we can get from no other materials.

The Roots of Meteoritics

That meteorites come from beyond the Earth is both a very old and a new idea. Through painstaking study of references to meteorites and related phenomena in the ancient literature of the Near East, Judith Bjorkman (1973) has shown that the Hittites accepted an extraterrestrial source at least 1,200 years B.C. They probably inherited the idea from still earlier inhabitants of the region, some of whom worked meteoritic iron at least 5,000 years ago. The ancient Greeks and Chinese also regarded meteorites as objects from the heavens, but this perception, like so much else of value, was lost to Western culture during the long intellectual night that we call the Dark Ages.

And it was recovered very slowly indeed. Although our meteorite collections include one stone, the Ensisheim chondrite, that fell almost 500 years ago, men long regarded that object and other "thunderstones" as products of volcanoes, storms, or other impressive but quite earthly phenomena. When they preserved newly fallen meteorites at all, it was usually for social or religious rather than scientific reasons. Thus Ensisheim, which fell in the Alsatian village of that name in 1492, found a place of honor in the village church only because Emperor Maximilian regarded it as an omen of victory against the Turks. If the Turks had beaten Maximilian, the history of meteorite recovery would be far shorter than it is, for two and a half centuries passed before another fall was preserved.

Although several important meteorite falls were recovered and described during the second half of the eighteenth century, the few men

who suggested that they came from beyond the Earth were either ridiculed or ignored.* This reaction was partly one of scientific snobbishness (meteorites usually fell at the feet of ignorant laymen whose testimony was not credible), but it also reflected a legitimate problem: there was no obvious connection between the stony meteorites that proceeded from large meteors and the irons and stony-irons that were unearthed as finds. The stony meteorites contained familiar minerals and looked enough like common rocks to make a terrestrial origin plausible. Though the irons and stony-irons were very different from known rocks and thus hard to explain by earthly processes, none of these meteorites had been seen to fall.

The idea of an extraterrestrial source of meteorites passed from fancy to fashion between 1790 and 1810, but historians differ on just when the change took place and who was responsible for it. Earlier accounts of the birth of meteoritics gave most of the credit to either E. F. F. Chladni or Jean-Baptiste Biot,† the former for a book written in 1794 that championed extraterrestrial sources of meteorites, the latter for a detailed and vivid account of the fall of the L'Aigle chondrite in 1803. In a recent review of the question, Derek Sears (1975) makes a strong case that Chladni's book appeared too early to account for the shift of opinion on the sources of meteorites, since Chladni, like his predecessors, was unable to establish the crucial link between iron and stony meteorites.

Sears shows that this vital step was taken in 1802, when Edward Charles Howard, a young English chemist, analyzed four chondrites at the request of the president of Great Britain's Royal Society. Howard showed that the four meteorites have almost identical compositions, though they fell in regions—England, Italy, India, and what is now Czechoslovakia—that are underlain by very different kinds of rock. That fact alone suggested to Howard that the four chondrites came from beyond the Earth, but his analyses of metal grains separated from them provided even stronger evidence. Howard found that the metal in chondrites contains nickel, an element that is also abundant in iron meteorites but absent from most terrestrial deposits of metallic iron. The missing link between stones and irons was missing no longer.

The collaboration between Howard and his colleague, de Bournon,

---

* D. Troili, who described the Albareto, Italy, chondrite in 1766, fared better than most such pioneers. We remember him in the name of the common meteoritic iron sulfide *troilite.*

† Chladni is remembered in *chladnite,* an archaic name for one type of enstatite achondrite. Mineralogists honor Biot in the name of the common black mica *biotite.*

produced many other important discoveries—the identification of troilite, for example, and the first description of chondrules—but we remember these men most for establishing that iron and stony meteorites are related. Sears argues, convincingly, that modern meteoritics was born in 1802, the robust child of a particularly fruitful marriage of chemistry and mineralogy.

This rereading of history diminishes the contribution of Biot, who has often been called "the father of meteoritics." Sent to L'Aigle in 1803 by the French Minister of the Interior, Biot returned with a vidid description of the L'Aigle fall that included testimony from a member of the Academy of Sciences and did much to persuade the Academy that meteorites come from space. Although that contribution was important—then, as now, the scientific establishment could either nurture a new idea or kill it—it was far less significant than Howard's discoveries of the previous year.

Skepticism about the origin of meteorites lingered well into the nineteenth century (Thomas Jefferson is said to have greeted a report of the 1807 fall of the Weston, Connecticut, meteorite with the comment, "It is easier to believe that Yankee professors would lie than that stones would fall from heaven"), but most scientists quickly accepted the conclusion that meteorites come from beyond the Earth. More important from the viewpoint of modern researchers, they set about the vital job of collecting and preserving these valuable objects.

Source Objects and Parent Bodies

Of the many conceivable extraterrestrial sources of meteorites, the moon seemed at first to be the most likely. It is close to the Earth, and as the Apollo astronauts demonstrated its modest gravity field (only one-sixth as strong as the Earth's) makes it relatively easy to lift heavy objects from its surface. Moreover, the moon seemed to have a ready source of energy to hurl rocks into space, for Sir John Herschel had reported that its surface features included active volcanoes. Though Herschel's report was later shown to be wrong—the moon's volcanic activity ceased about 3 billion years ago—many influential scientists, including Biot, Poisson, and Laplace, accepted a lunar origin for meteorites during the early 1800s.

Asteroids, the first and largest of which (Ceres) was discovered in 1801, quickly joined the moon as promising sources of meteorites. Since most of these small bodies have orbits in a region of the solar system that lacks a planet, it seemed plausible that asteroids are the remains of such an object and that meteorites in turn are bits of the asteroids.

The idea that meteorites of all kinds come from one planetary *parent body* by way of asteroidal *source objects* took root early in the nineteenth century and held sway for more than 150 years. Geologists found the idea of a "tenth planet" particularly attractive because it held out the hope that they could learn about the Earth's hidden interior by studying the dissected remains of one of its siblings. In fact, one geologist, Reginald Daly, constructed in 1943 a detailed model of the meteorite parent planet, which he pictured as a bit smaller than Mars and composed of an achondritic crust, a chondritic mantle, and a core composed of iron meteorites.

As we shall see in the next chapter, the idea that all meteorites formed in one big parent body is no longer acceptable. Moreover, we now know that where meteorites come from and where they formed are really two separate questions, and we use different tools to answer them. The remainder of this chapter will be devoted to the first question: where are meteorites coming from today?

Geography of the Solar System

We are sure that meteorites come from beyond the Earth and almost as certain that they come from within the solar system. Although samples from beyond the reach of the sun's gravitational field would be very exciting, no meteorite now in hand is unusual enough in chemical or isotopic composition to suggest that it comes from outside the solar system. Indeed, the extremely high velocities of such extrasolar objects make it unlikely that any would survive passage through the Earth's atmosphere.

Before we can choose among the many possible sources of meteorites within the sun's family, we need to know what and where they are and how they move around the sun. We must, in other words, know something about the geography of the solar system.

Planets. In an age when we can see a rocket lift off from Cape Canaveral one morning and watch its passengers hop about on the moon a few days later, it is easy to picture the solar system as a cozy little neighborhood. Even the fact that it takes sunlight 5½ hours to reach Pluto is likely to impress us less than the ease with which television's starship *Enterprise* scuttles from galaxy to galaxy between commercials. To restore our sense of the vastness and emptiness of space,* we need to scale the solar system down to familiar

---

* It is so immense that distances expressed in miles or kilometers are cumbersome. We use instead the *astronomical unit* (a.u.), which is defined as the distance between the earth and the sun, about 93 million miles.

terms. Suppose that the nine planets were laid out on a football field, with the sun on one goal line, Pluto on the other, and the other eight planets spaced according to their relative distances from the sun (Figure 5.1a). What would this miniature solar system look like from a choice seat on the 50-yard line?

It would look like an empty football field! If you had binoculars, you might spot a glowing 1-inch sphere—the sun—on one goal line, but you would see none of its relatives. For Jupiter, the largest planet, would be no bigger than a dried pea, and the Earth, huddled with Mercury, Venus, and Mars inside the sun's 5-yard line, would be smaller than a grain of sand.

This drawing makes some useful points beyond the obvious one that the solar system is immense and nearly empty. For example, it shows that the planets are not distributed randomly around the sun but fall into two groups. The four *inner* or *terrestrial* planets are small bodies that lie close together near the sun; the *outer* or *giant* planets are much larger, far more distant, and much more widely spaced (Table 5.1). Members of these two groups differ in other respects as

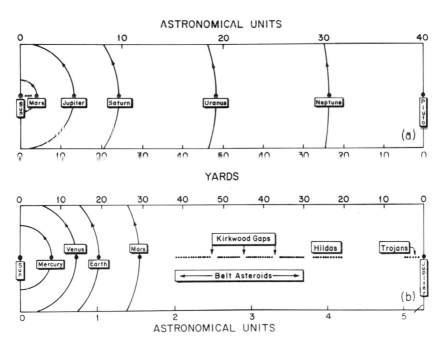

Figure 5.1 The solar system in a football stadium. (*a*) The sun is located on the left-hand goal line, Pluto on the right, and the other planets are spaced proportionally between them. (*b*) Detailed view of the region between the sun and Jupiter, showing the orbits of the terrestrial planets and asteroids.

Table 5.1 Vital statistics for the sun, planets, and the moon

Body	Radius (mi.)	G	Gravity
Sun	432,443	1.4	27.9
Mercury	1,512	5.4	0.4
Venus	3,761	5.3	0.9
Earth	3,959	5.5	1.0
Moon	1,080	3.3	0.2
Mars	2,100	3.9	0.4
Jupiter	43,346	1.3	2.7
Saturn	36,176	0.7	1.1
Uranus	14,584	1.6	1.1
Neptune	14,115	2.3	1.4
Pluto	3,542	1.7	0.2

Note: Specific gravity (G) is the weight of a body divided by the weight of an equal volume of water; for comparison, common silicate minerals have G's between 2.5 and 3.5, and metallic iron has a G of about 7.5. The gravitational acceleration for each body is given relative to the Earth's.

well. For example, the four terrestrial planets and the moon are dense bodies, whose specific gravities lie between those of common silicate minerals and that of metallic iron, as shown in the table. Even if we had not sampled the moon and sent landers to Mars and Venus, we would conclude from this that they must be rocky objects and probably contain more or less metallic iron as well. On the other hand, the low specific gravities of the other planets and most of their satellites suggest that they contain little rocky material but instead consist mostly of gases or ices of light compounds such as water, ammonia, carbon dioxide, and methane.

Figure 5.1a implies that the planets move around the sun in the same direction (counterclockwise as viewed from the north), in the same plane, and at fixed distances from the sun. The first implication is correct and applies to most, though not all, satellites as well. The second and third implications are only approximately correct.

The planets and all other bodies that are gravitationally bound to the sun follow elliptical orbits, with the sun at one focus of each ellipse (Figure 5.2). Thus, such bodies are not always at the same distance from the sun but move between two extremes that are called *perihelion* (nearest to the sun) and *aphelion* (farthest from it). As the examples in Figure 5.2a show, the elliptical orbits of bodies in the solar system vary widely in form, from very elongate or *eccentric* to almost circular. Most of the planets have almost circular orbits: their perihelia and aphelia differ by 10 percent or less. The orbits of the smaller planets—Mercury, Mars, and Pluto—are more eccentric. In fact, the difference between Pluto's perihelion and aphelion is so large (40 percent) that this planet is sometimes closer to the sun than Neptune is.

Planetary orbits also lie in different planes and thus cross the plane of the Earth's motion—the *ecliptic*—at various angles (Figure 5.2b). Most of the planets have very low angles of inclination (from less than 1 degree for Uranus to 7 degrees for Mercury), but Pluto is again an exception. Its 17-degree inclination, its eccentric orbit, and its unusually small size relative to the other outer planets lead astrono-

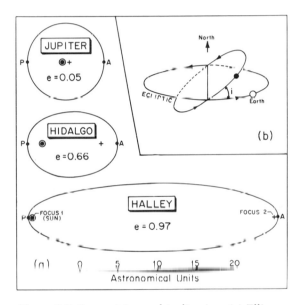

Figure 5.2 Eccentricity and inclination. (*a*) Elliptical orbits of a planet (Jupiter), an asteroid (Hidalgo), and a periodic comet (Halley). Each diagram shows the sun (circle with dot) at one focus; the other focus is shown by a cross, and perihelion and aphelion are indicated by P and A, respectively. The eccentricity (e) increases from Jupiter to Hidalgo to Halley. (*b*) Orbits of the Earth and another object moving in the same sense around the sun. The angle between the two orbital planes is the inclination (i) of the second object. Where i exceeds 90 degrees, the orbit of the second body is retrograde.

mers to wonder whether Pluto is really a planet at all. Was it always where it is now, or is it an escaped satellite of another planet, possibly Neptune?

In summary, then, the rocky inner planets and the icy and gaseous outer ones move around the sun in the same direction and in almost circular orbits that lie in nearly the same plane. Even Pluto, though something of a maverick, is much less so than some of the smaller members of the sun's family.

Asteroids. Our solar system in a stadium shows how the largest members of the sun's family are distributed, but it tells us nothing about the smallest ones: comets and asteroids. In both cases, the problem is one of scale. Many comets approach the solar system from sources that are thought to be 10 to 100 *thousand* astronomical units (a.u.) from the sun. If we included comets and their distant source region in Figure 5.1, all nine planets would shrink to form one tiny speck on the sun's goal line.

The asteroids also raise a problem of scale, for most of them occupy a tiny region in Figure 5.1a. To see how these bodies are distributed relative to the sun, we need to enlarge the picture to focus on the inner solar system. Figure 5.1b does this by putting Jupiter instead of Pluto on the goal line and redistributing the other planets accordingly. This step improves our view of the planets only slightly—the sun is still only 6 inches in diameter, Jupiter about half an inch across— but it allows us to look in some detail at the "71-yard" gap between Mars and Jupiter. This gap contains most of the small, rocky objects that we call asteroids. As Figure 5.1b shows, these bodies, the largest of which (Ceres) is less than 500 miles in diameter, are not distributed uniformly between Mars and Jupiter: most of them move in a band between 2.1 and 3.7 a.u. from the sun. This band is known as the asteroid belt, and its inhabitants are called the *belt* or *ring* asteroids. Though most asteroids move around the sun within this main belt, many do not. Two groups, the Hildas and the Trojans, have orbits that carry them closer to Jupiter; a few of the Trojans in fact go beyond that planet. At the other extreme, a few bodies, called Amor asteroids, approach the Earth's orbit and some, the Apollo asteroids, cross it (see Figure 5.3). These Earth-approaching and Earth-crossing asteroids are obviously prime candidates for meteorite source objects.

A curious feature of the asteroid belt, also evident in Figure 5.1b, is that its occupants are not distributed randomly but are concentrated in four narrower belts that are separated by three almost asteroid-free zones. These empty zones, called *Kirkwood gaps,* occur at distances from the sun such that objects within them take one-third, two-fifths,

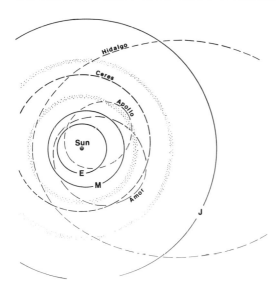

Figure 5.3 Orbits of the terrestrial planets and asteroids. Stippled rings outline the asteroid belt, one inhabitant of which (Ceres) is shown separately. Also shown are the orbits of a Mars-crossing asteroid (Amor), an Earth-crosser (Apollo), and one asteroid (Hidalgo) whose very eccentric orbit carries it far beyond Jupiter.

or one-half as long to revolve around the sun as does Jupiter. These relationships make orbits within the gaps unstable: an object that enters a gap quickly leaves it on an eccentric orbit that may carry it into the inner solar system. The Kirkwood gaps appear to be an important link in the chain of events that delivers meteorites to the Earth.

Asteroids move in the same direction as the planets but in varied orbits (Figure 5.3). The belt asteroids (Ceres, for example) have orbits that are only slightly eccentric and, in most cases, only modestly inclined to the ecliptic. That these asteroids move around the sun in much the same way as their larger neighbors justifies their alternate name, minor planets. The Mars-crossing and Earth-crossing Amor and Apollo asteroids have more eccentric orbits, but they too cross the ecliptic at low angles of inclination. Some other asteroids are much less well behaved. For example, Pallas, the second largest asteroid (about 300 miles in diameter), has an inclination of 37 degrees, and tiny Hidalgo (23 miles in diameter) crosses the ecliptic at 42.5 degrees as it shuttles back and forth from well beyond Jupiter to slightly beyond Mars.

Comets. As described thus far, the solar system is a rather orderly arrangement of objects that move in the same direction around the sun, in orbits of slight to moderate eccentricity and low inclination. Comets are far less orderly, for although a few of them move in the same region and in much the same way as many asteroids, most

approach the sun from far greater distances and on extremely elongate orbits.

Spectacular photographs of a few unusually handsome comets lead
us to think of these objects as huge, fiery bodies with long, glowing
tails (Figure 5.4). In fact, the solid heads or *nuclei* of comets are very
modest objects—rarely more than 10 miles in diameter, typically
much smaller—composed of a mixture of ice and dust that has been
described, perhaps accurately but unappealingly, as being like a
"dirty snowball." Even the brilliant light show that some comets
produce when they pass near the sun does not always appear, for it
depends on the effect of charged solar particles—the solar wind—on
ices in the nucleus. If a comet's nucleus is ice-poor, it will produce a
modest show or none. Comet Kohoutek, which appeared a few years
ago, was disappointingly tame, perhaps for this reason.

Comets have varied orbits that reflect various histories. Perhaps the
best way to approach them is by describing the most celebrated one
and comparing it with others that are less well known but more
important to our inquiry into the sources of meteorites. The most
famous of all comets was discovered by and is named for Edmund
Halley, who calculated its orbit in 1682. Though Halley at first
thought its orbit had the form of a parabola, he noted that it re-
sembled orbits calculated for comets that had appeared in 1531 and
1607. He concluded from this similarity that what seemed to be three
different comets were really three appearances of one object, which

Figure 5.4 Halley's Comet, as photographed in 1910 in Chile. (Lick Observatory photograph, reproduced with permission.)

was traveling in a very elongate elliptical orbit. From the intervals between its earlier appearances Halley predicted that the comet would appear again in 1758, and it did: on Christmas night.

On the basis of historical records that extend back to 240 B.C., Halley's comet has appeared at least 28 times at an average interval of 77 years. It appeared in 1066, when the Normans conquered England, and it is shown in the Bayeux tapestry. It also lit the sky when Samuel L. Clemens was born in 1835 and returned when, as Mark Twain, he died in 1910. In late 1985 and early 1986, during its first visit to the inner solar system since the dawn of the space age, Halley's comet was greeted by a small fleet of European and Japanese spacecraft and studied by eager professional and amateur astronomers around the world.

Because Halley's comet returns to the inner solar system at regular intervals, we call it a *periodic* comet. Its orbit is very eccentric, with an aphelion slightly beyond Neptune, and its inclination is so great—more than 90 degrees (see Figure 5.2b)—that it moves around the sun in a retrograde sense, that is, clockwise as viewed from the north.

Other comets whose aphelia are smaller than Halley's are also periodic. Some of them, for example Encke's comet, have aphelia that lie inside the orbit of Jupiter. Most such short-period comets have direct orbits of low inclination that resemble quite closely the orbits of some asteroids. Their nuclei appear to be ice-poor, for these objects glow only modestly as they pass the sun.

Most comets range much farther from the sun than Halley's comet, and, unlike the other objects that have been discussed, they approach the sun from every direction. Though we call such objects *long-period* comets, we are not sure whether they are really periodic or not. Measurements of their trajectories is usually when they are near the sun, cannot distinguish between closed (elliptical) or open (parabolic) orbits, and historical records of comets are of little help to us for objects that return at intervals of several hundred years, if they return at all.

Whether the far-ranging comets are or are not periodic is of less concern to meteoriticists than is evidence that these objects are related to the short-period comets that have asteroid-like orbits. Just how comets pass from one type of orbit to the other is unclear, but there are two attractive possibilities. One, suggested by the fact that several well-known comets have aphelia close to Jupiter, is that these objects—sometimes called "Jupiter's family"—were captured by that planet's gravitational field. The other interpretation, which is more popular today, is that short-period comets are actually long-period comets that have lost much of their ice during repeated encounters with the sun. At each such encounter, the solar wind drives off some ice and modifies the orbit of the remaining material. Though a comet's mass and orbit change only slightly during one pass by the sun, the changes wrought by many such passes are dramatic: in a period of 10,000 years or so, the comet's nucleus shrinks from a mass of ice and dust to one made up largely of dust, and its orbit changes from highly eccentric to more nearly circular.

That Comet Encke and its cohorts appear to have ice-poor nuclei is a point in favor of this second interpretation. Thus some of the smaller bodies that we categorize as asteroids and assume were always members of the inner solar system may really be exhausted comets, spending their retirement close to the warm sun after a lifetime spent commuting from cold space far beyond it.

Identifying Meteorite Source Objects

Obviously, the sun's family includes a great many objects, any or all of which might conceivably launch samples toward the Earth. How can we identify those that did? The first step is to eliminate those bodies that, for reasons of composition or history, cannot be meteorite source objects; as we shall see, this step narrows the field dramatically. The second step is to compare the orbits of meteors and recovered meteorites with those of possible source objects, and the third is to see whether other lines of evidence confirm or contradict conclusions drawn from orbital data.

Clearly meteorites do not come from objects that consist chiefly of gas or ices; hence we can at once eliminate the outer planets and most of their satellites as possible source objects. Differentiated meteorites, some achondrites in particular, might come from the rocky inner planets, but such sources are most unlikely for chondrites. Data from unmanned landings on Mars and Venus show that these planets had long histories of internal melting and volcanism that almost certainly destroyed all vestiges of the primitive materials from which they formed. Though we know Mercury only from photographs and fly-bys, this conclusion probably applies to that planet as well.

Before we visited and sampled it, the moon seemed to be a likely source of some kinds of meteorites, in particular those achondrites that resemble terrestrial volcanic rocks. But the Apollo and Luna programs effectively put an end to that idea. With just four exceptions, meteorites resemble lunar rocks only superficially and differ from them in ways (age, isotopic composition) that prove conclusively that they are unrelated.

The exceptions are, however, very interesting, for they bear on whether some meteorites might indeed come from another planet. In 1982, researchers who were hunting for meteorites in Antarctica picked up a small, odd-looking rock that proved on closer study to be a piece of lunar breccia. To the great surprise of those who studied it, this breccia apparently experienced only modest shock damage when an impact ejected it from the moon. This observation is very important, for most meteoriticists have assumed that an impact violent enough to drive samples from a body as big as the moon would at least shock them beyond recognition, if it did not melt or vaporize them. Evidence that planetary bodies can yield objects with modest shock histories removes an important stumbling block from the path of the exciting idea that our meteorite collections include a few samples from Mars (see Chapter 9).

If the concept of comet nuclei as "dirty snowballs" is correct, these objects are not promising sources of meteorites. Their small sizes and the presence of ices make it inconceivable that comet nuclei ever experienced temperatures high enough to metamorphose chondritic material or melt it to produce differentiated meteorites. Of all the meteorites known to us, only a handful of carbonaceous chondrites—those that lack chondrules—had low temperature histories that are consistent with an origin in comet nuclei. And even these meteorites bear evidence that makes a cometary source most unlikely.

When we eliminate those bodies that cannot be sources of meteorites, we are forced to conclude that most and probably all meteorites come from sources within the inner solar system. Meteoroid orbits point to the same conclusion. Photographically determined orbits for three recovered meteorites indicate that they approached the Earth in orbits of low inclination with aphelia inside or just beyond the orbit of Jupiter (Figure 5.5; compare Figure 5.3). Though this evidence is meager and limited to ordinary chondrites, it is consistent with the far more numerous orbits calculated for fireballs, the brilliant meteors whose motions indicate that they are produced by dense rocky or metallic meteoroids. It seems clear that most dense meteoroids, and hence most or all meteorites, come from somewhere between Mars and Jupiter—that is, from the asteroids, short-period comets, or, conceivably, Mars.

Though the orbits of fireballs resemble those of recovered meteorites, those of the smaller meteors that we call shooting stars are far more varied. Some of the tiny, fluffy meteoroids that produce such meteors approach the sun from random directions that recall the orbits of long-period comets. Others travel in clusters or *streams* that

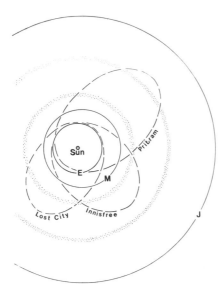

Figure 5.5 Photographically determined orbits for three recovered meteorite falls, the inner planets, and the belt asteroids.

create meteor showers when they enter the Earth's atmosphere. Most such streams either appear with or are in the same orbits as known periodic comets. Thus, for example, the Taurid stream that lights the sky each fall is associated with Encke's comet, and the Draconid stream, which has produced the most vivid meteor showers observed in this century, travels with another short-period comet, Giacobini-Zinner.

Thus it is clear that meteor streams proceed from comets, and it is likely that the process that forms such streams eventually destroys those comets by removing the ice that holds their nuclei together. The experience of Biela's comet may show us how comets die. This object returned every 6½ years until 1846, when it appeared as two bodies traveling together. The two parts of Biela appeared just once more, much farther apart. Then the comet vanished, to be replaced by a stream of meteors in the same orbit: the Andromedids or Bielids.

If short-period comets like Encke, Giacobini-Zinner, and Biela yield meteors, might they not yield carbonaceous chondrites as well? A subtle but powerful argument against that possibility concerns the solar wind that was mentioned in connection with comets and their glowing tails. This stream of solar particles not only strips ices from comet nuclei but also damages rocks that are exposed to it and implants noble gases (for example, helium, argon, and xenon) of distinctive isotopic composition. As Edward Anders has observed, meteorites that contain large amounts of such solar-type noble gases must have spent very long periods of time within a few astronomical units of the sun—far longer than the estimated 10,000-year lifetime of a short-period comet. Since most meteorite classes, including the carbonaceous chondrites, include gas-rich members, this argument makes it most unlikely that any meteorites proceed from comets. It appears rather that all meteorites come from source objects that are now, and always were, members of the inner solar system.

My arguments concerning the sources of meteorites have thus far been largely negative. Though scientists often have to rely on such arguments, we are much happier when we can prove hypothesis A than we are when we can only disprove B, C, and all other alternatives that we can think of. For it is always possible that some of our negative evidence will prove to be false (like the argument that planetary ejecta must be severely shocked), or that another explanation—hypothesis X—will occur to us at a later time. Is there any *positive* evidence, other than orbital data, that most or all kinds of meteorites come to the Earth from asteroids? There is such evidence, and it comes from just looking at the two kinds of objects—though, admittedly, in a very special way.

In the last two decades, comparisons of sunlight reflected from the surfaces of asteroids and that reflected from samples of meteorites have been used to determine whether these objects consist of the same minerals. The basis of this approach, which is called *reflectance spectrophotometry,* is that most minerals reflect light at some wavelengths and absorb light at others. Thus, for example, a ruby looks red to us because it absorbs light in that part of the visible spectrum that we associate with the complementary color, green; an emerald, on the other hand, looks green because it absorbs red light strongly. A spectrophotometer, which measures the ratio of reflected to incident light—the sample's *albedo*—at various wavelengths in the visible and infrared spectrum, can make much subtler distinctions than we can make with our unaided eyes. This instrument yields a pattern of reflectance versus wavelength that is characteristic of one mineral or a particular combination of minerals (Figure 5.6). Although the total reflectance of a sample—whether it appears light or dark—depends on its texture (coarse-grained, fine-grained, or powdery), the shape of its spectrum varies far less and is nearly as distinctive as a human fingerprint.

Reflectance spectra have been measured for almost all kinds of meteorites and for about 100 asteroids. The results of this work strongly support the conclusion that meteorites come from asteroids, for each kind of meteorite has a spectral equivalent in one or more

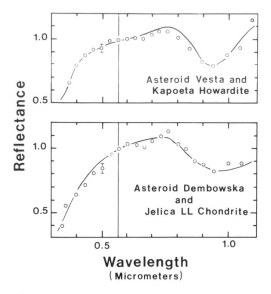

Figure 5.6 Comparison of the reflectance spectra of two asteroids (solid lines) and two meteorites (circles). (Data are from Chapman, 1976).

asteroids. In some cases, for example that of some basalt-like achondrites and asteroid Vesta (Figure 5.6), the uniqueness of an asteroid's spectrum suggests that it is the sole source of a particular kind of meteorite. In other cases, carbonaceous chondrites for example, there seem to be a great many possible source asteroids. In fact, one of the most important conclusions from the study of asteroid and meteorite spectra is that the Earth receives a very biased sample of the meteorite source objects. Though there are relatively few carbonaceous chondrites in our collections, carbonaceous asteroids appear to be very numerous. On the other hand, surprisingly few asteroids have spectra that match ordinary chondrites, which are by far the most common of meteorites. Obviously, meteorite delivery is an inefficient business in which chance plays a very large role.

Meteorite Delivery

The conclusion that most or all kinds of meteorites come from asteroids raises a problem. Clearly, the bodies that deliver meteorites to the Earth must move in paths that intersect the Earth's orbit, yet most asteroids—those that populate the asteroid belt—move in nearly circular orbits whose perihelia lie well beyond Mars. Moreover, spectral reflectance data for belt asteroids suggest that their orbits are very stable: asteroids of different composition tend to occur in different parts of the belt, implying that there has been very little mixing across the belt for billions of years. How can pieces of such stable objects find their way to Earth?

The Earth-crossing and Mars-crossing Apollo and Amor objects (see Figure 5.3) appear to answer this question, but in fact they just restate it in different terms. Obviously such asteroids are well positioned to yield meteorites, and we are confident that they do so. However, they are at best temporary sources, for bodies that cross the paths of planets tend to be short-lived: they are vulnerable to collisions that would destroy them outright and to gravitational perturbations that would change their orbits, sending them into the sun or out of the solar system. Thus the cast of Earth-crossing asteroids must change with time. The question we have to answer is how that cast is replenished with new material from the asteroid belt.

One important clue to the mechanism of meteorite delivery is its rate—the time it takes for meteorites to make their way from the asteroid belt to the Earth. We can determine this by calculating how long they have been exposed to cosmic rays, that is, their *cosmic-ray exposure* or *CRE ages*. The basis for such ages is the fact that cosmic rays interact with the elements in meteorites to produce distinctive isotopes. I mentioned one such isotope, aluminum-26, in Chapter 1 in connection with attempts to measure the Earth's harvest of meteoritic material; others that are used for determining CRE ages are helium-3, neon-21, and argon-38. Since a modest covering of metal or rock suffices to shield a sample from cosmic rays, the abundances of such *cosmogenic* isotopes can be used to calculate when a meteorite was liberated from its parent body to form an object with a diameter of a few feet or less.

Exposure ages vary widely and more or less regularly with meteorite type. For example, most iron meteorites yield ages between 200,000 and a billion years, while most chondrites have ages of less than 60 million years. Though such ages suggest that a meteorite's journey from the asteroid belt to the Earth is very leisurely indeed, they are really surprisingly short. Calculations made in the 1960s showed that the exposure ages measured for most chondrites allow too little time for random gravitational perturbations to move a belt asteroid into an Earth-crossing orbit. Massive collisions between asteroids would speed up the process, but such collisions are neither frequent in the belt nor are they recorded in most chondrites. In the late 1960s and early 1970s, the problem of getting material from belt asteroids to the Earth on time seemed to be insurmountable, a fact that led several researchers to reject asteroidal source objects in favor of comets or the moon.

We now believe that the key to swift delivery of meteorites lies in the Kirkwood gaps, zones of orbital instability within the asteroid belt. Belt asteroids are nudged into a gap by gravitational perturbations or gentle impacts. The strong gravitational accelerations that the asteroids experience in the gap throw them into highly eccentric paths, some of which then evolve into Earth-crossing orbits. Though we understand the processes involved in this mechanism in only a general way, it seems to work, and on a time scale that agrees with the exposure ages. Thus we think we know not only where most meteorites come from but how.

Before we leave the subject of meteorite source objects, it is worth taking note of an important implication of this picture of meteorite delivery. If the present source objects are temporary, it follows that different objects yielded those meteorites that struck the Earth thousands or millions of years ago. The possibility that some of those objects were very different from modern Earth-crossing asteroids is an important reason for our intense interest in the ancient meteorites preserved in the Antarctic ice sheet.

6

Chondrite Parent Bodies

I t is hard today to imagine the uproar that greeted Robert Fish, Gordon Goles, and Edward Anders when they suggested that meteorites both come from and formed in asteroids. Although a Soviet scientist, Boris Levin, had made a similar suggestion in 1958, it was a paper that these three men from the University of Chicago published in 1960 that introduced the idea to Western scientists and touched off the second revolution in meteoritics.

And a lively revolution it was, fiery enough to shock those of us who had just emerged from the snug cocoon of graduate school and imagined scientists to be patient, gentle seekers of truth. I recall a particularly stormy meeting in 1963 or 1964, where a prominent "large-body person" presented a paper and an eminent "small-body person" challenged it. The two men thundered at each other for 10 minutes, while the moderator tried in vain to make peace and move along to the next speaker.

No one who has read James Watson's book *The Double Helix* will be surprised to find that scientists are a proud, passionate lot. But why was there such a fuss over this particular issue? Why would a question as arcane as the sizes of meteorite parent bodies lead men of normally good will to argue so vehemently? The answer is that these scientists realized that the question held the key to the significance of meteorites and that the answer to it would determine their role in planetary research. They knew that the Earth's long, complex history of melting and igneous activity had scrubbed away all vestiges of the material from which it formed, and they reasoned that this was likely to be true of other planets as well. Samples of such bodies might help

69

us determine how the Earth and the other planets evolved, but they would probably tell us very little about how, and from what, they formed.

Samples of preplanetary material were far more likely to survive in smaller bodies—objects that were too small to retain much radioactive heat and thus had experienced short, simple igneous histories or none at all. One important scientific reason for man's historic race to the moon was the hope that it had preserved some evidence of its origins. And one disappointing result of that race was the discovery that it had not: even though the moon's igneous history ended 3 billion years ago, it was violent enough to destroy all direct evidence of our satellite's beginnings.

Thus the question of parent body size was crucial. If all such bodies were big and Earth-like, meteoritics would continue to be what it had been before: a modest handmaiden of geology. If, on the other hand, some of the parent bodies were small and undifferentiated, meteoritics would move to the front rank in the search for the solar system's beginnings.

A second reason for the passionate argument about this issue in the early 1960s was that evidence for the sizes of parent bodies was sparse and contradictory. Proponents of large parent bodies argued that a small body could not have internal pressures that were high enough to produce the diamonds present in the Canyon Diablo iron meteorite and in the small group of achondrites known as ureilites. Nor could a small body hold enough of its meager radioactive heat to cause the metamorphism recorded in most chondrites and the melting that produced differentiated meteorites. Finally, if such a body did manage to melt, its gravitational field would be too feeble to cause molten metal to sink to form a core, too weak to make crystals sink or float to produce the various kinds of achondrites.

Although these arguments struck most researchers of the time as compelling, the proponents of small bodies raised counterarguments that seemed equally strong. They pointed out, for example, that if all the material in the asteroid belt were gathered together, the result would be a body smaller than the moon. Even if the belt held more material 4.5 billion years ago, it was unlikely to have been enough to make one planet, much less the several bodies suggested by the sharp chemical differences among various types of meteorites. Another argument against the view that asteroids are bits of a shattered planet was that most such bodies are round rather than jagged; a third was that few meteorites show evidence of the intense shock that should have accompanied destruction of a planet.

Advocates of small parent bodies put forward a fourth argument that took the heat problem raised by the large-body camp and turned it upside down. If it was hard to understand how small bodies could become hot enough for metamorphism and melting, it was even harder to see how a body as big as the moon could cool sufficiently to produce the Widmannstätten pattern observed in most iron meteorites (see Chapter 8). This argument shows vividly how equivocal the evidence for parent body sizes was in 1960, for it made one meteorite—Canyon Diablo—a star witness for both the defense and the prosecution: the diamonds in Canyon Diablo seemed to require a big body, its Widmannstätten pattern a small one.

Twenty-five years after the fact, we can see that most of the arguments on both sides of the issue of body size were weak. The high pressure signified by the presence of diamond in some meteorites does not necessarily testify to deep burial in a large body; the problem of heating small bodies has vanished with the discovery of heat sources other than long-lived radioactivity; and we know from experiments that processes of differentiation can go forward in surprisingly weak gravity fields. On the other side, the original mass of the asteroids was, and remains, poorly known; and, as noted in the previous chapter, the discovery of modestly shocked meteorites from the moon indicates that intense shock does not always accompany the ejection of material from a large body.

Given the inconclusiveness of the evidence used by both camps in the early 1960s, the question of body size might well have been debated for decades. It was, in fact, settled in less than five years because of a breakthrough in the field of nuclear chemistry on a separate but related question: that of the *number* of meteorite parent bodies.

How Many Parent Bodies?

Early proponents of small meteorite parent bodies recognized that the sizes of such bodies should be related to their number. If the mass of the asteroids were distributed among many bodies, most or all of them would have to be much smaller than the moon. In principle, we could determine the size of an average parent body by dividing the original mass of the asteroids by the number of parent bodies. Unfortunately, the numerator of this equation is very uncertain. The present low rate of loss of material from the asteroid belt suggests that that region never contained much more mass than it does today, but we are by no means sure that the rate was constant in the distant past.

Recent calculations suggest that the original mass of the asteroids was about twice that of the moon, but it could have been as large as, or even larger than, the mass of Mars.

Though the numerator of this equation remains elusive, we have made considerable headway with the denominator, for we have learned how to determine which meteorites are or are not closely related and thus how many parent bodies meteorites represent. In Chapter 4 we saw that chondritic meteorites fall into eight or nine chemically distinct clusters. That the various groups of chondrites do not intergrade (there are, for example, no meteorites that lie midway between carbonaceous and ordinary chondrites) suggests that they formed far apart. Chondritic breccias point toward the same conclusion, for even though the fragments that make up such breccias record varied metamorphic and shock histories, most of those in a particular breccia have very similar chemical compositions. Thus we seldom find bits of enstatite chondrites mixed with bits of ordinary or carbonaceous chondrites.

These observations indicate that the eight or nine chondrite groups developed in almost complete isolation from one another, but does this prove that they formed in different bodies? We need look no further than the Earth and moon to see that it does not. The Earth's crust consists of rocks that differ much more widely in composition than chondrites, yet they obviously formed in one body. If the Earth's atmosphere were stripped away, repeated meteorite impacts would pound and mix these varied rocks as they have mingled fragments on the airless surface of the moon. But even on the moon, most breccias consist very largely of local materials.

Obviously neither sharp chemical differences among chondrites nor the prevalence of one-group breccias suffices to prove that chondrites come from more than one body. In fact, these arguments work much better in reverse: chemical intergradations among several kinds of achondrites and their frequent association in breccias are powerful evidence that these meteorites formed close together, probably in one parent body. To prove that two groups of meteorites formed in different parent bodies, we need stronger evidence than was available in 1960. Fortunately, stable isotope geochemistry has provided just such evidence.

We have already seen that oxygen consists of several isotopes, the most abundant of which have atomic weights of 16, 17, and 18 mass units. Geologists have known for many years that chemical and physical processes work to separate the most abundant oxygen isotope, O-16, from the heavier isotopes. For example, when a clam takes calcium carbonate from seawater and adds it to its shell, the proportions of O-16, O-17, and O-18 in the shell differ slightly from those in seawater. Since the extent of these differences varies with temperature, marine geochemists can use the oxygen isotopic compositions of shells of various ages to study changes in seawater temperature during and between ice ages. Other geochemists use the same principle to determine the temperatures of formation of igneous and metamorphic rocks and ores.

Our long experience with oxygen isotopes in rocks has shown us that O-17 and O-18 do not vary at random relative to O-16 but follow a very specific pattern. Since O-17 is one atomic mass unit heavier than O-16 and O-18 is two units heavier, a process that enriches or depletes the heavy isotopes relative to O-16—a mass-fractionation process—has twice the effect on O-18 that it has on O-17. Thus a clam shell that contains one part per thousand (per mil, shown symbolically as 0/00) more O-17 than seawater will contain two parts more O-18. In other words, the two heavy isotopes are enriched or depleted in a 1:2 ratio.

This relationship is shown in Figure 6.1, which compares the 17/16

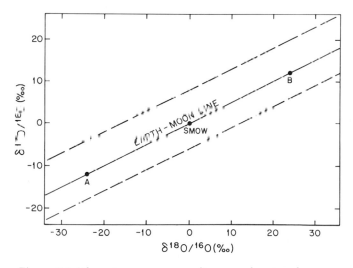

Figure 6.1 Three-isotope oxygen diagram, showing the Earth-moon mass-fractionation line (solid line) and two samples on it. Dashed lines show mass-fractionation lines for samples from different oxygen reservoirs. Isotopic ratios (17/16 and 18/16) in samples are expressed as deviations (deltas) from those found in standard mean ocean water (SMOW).

and 18/16 ratios in two minerals with the same ratios in standard ocean water. Sample A has a 17/16 ratio that is 12 per mil smaller than the standard ratio and an 18/16 ratio that is 24 per mil smaller. This sample and sample B, which shows similar enrichments in the heavy isotopes, lie on a line through the standard with a slope of 1/2.

We might expect all samples that formed from the same oxygen reservoir to lie on the same line, and they do: all terrestrial materials lie on the solid line in Figure 6.1. So do all lunar samples, indicating that the Earth and moon formed from the same oxygen reservoir. This fact is strong evidence that our planet and its satellite formed in the same region and are not chance associates from different parts of the solar system. If the moon came from a different oxygen reservoir, lunar samples would fall on a different mass-fractionation line, perhaps one of the dashed lines in Figure 6.1.

In the early 1960s we had no idea whether oxygen has the same isotopic composition throughout the solar system or varies from one object to the next. This question was answered in 1965, when Hugh Taylor and his co-workers at the California Institute of Technology published oxygen isotopic analyses of several kinds of meteorites. Some of these meteorites proved to have similar isotope ratios, suggesting that they formed in the same oxygen reservoir. Others differ enough in isotopic composition to make it clear that they formed in different reservoirs.

Since Taylor's pioneering study, the distribution of oxygen isotopes has been examined in nearly all kinds of meteorites, and this has become a very important tool for reconstructing the limbs, branches, and twigs of the meteorite family tree. Figure 6.2 summarizes some results of this work, most of which has been done by Robert Clayton and his colleagues at the University of Chicago. This figure shows that the iron-rich (H) and iron-poor (L, LL) ordinary chondrites lie in different fields above the Earth–moon line. The enstatite chondrites and the closely related enstatite achondrites define another field that straddles the line, and several kinds of achondrites lie below it. The carbonaceous chondrites tell quite a different story, for they scatter along a line with almost a 1:1 slope. This pattern has exciting implications, which will be discussed further in Chapter 10; for the moment it suffices to note that carbonaceous chondrites have oxygen isotopic compositions that are quite different from those of other meteorites, providing strong evidence that they formed in one or more different reservoirs.

The pattern in Figure 6.2 is the strongest evidence we have that chondrites and other meteorites formed in many parent bodies rather

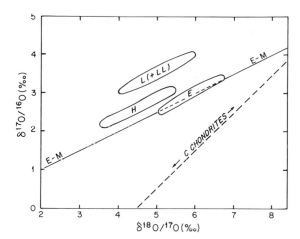

Figure 6.2 Three-isotope oxygen diagram for several kinds of chondrites (H, L, and LL ordinary; enstatite; carbonaceous), compared with the Earth-moon (E-M) line.

than one body. By using chemical, mineralogical, isotopic, and other kinds of evidence, meteoriticists have shown that the silicate-bearing meteorites and the major groups of irons sample about 20 bodies. The many iron meteorites that do not fit in the larger groups on chemical grounds may sample perhaps 50 more bodies, for a total of 70.

Though the lowest modern estimate for the number of meteorite parent bodies—20—would amaze our forebears, it is quite likely that even the highest—70—is too low. As we saw in the previous chapter, objects in the asteroid belt move in very stable orbits and leave those orbits with great reluctance. It is certain that many asteroids have not yet favored the Earth with samples, while others have made their contributions to the Earth and vanished, swept up by a planet or the sun or ejected from the solar system. That earlier generations of meteorites were quite different from those that fall today is one of the exciting possibilities that we can test through careful study of the ancient meteorites from Antarctica.

Sizes of Parent Bodies

The fact that there were at least 20 meteorite parent bodies and probably many more is strong evidence that most or all of those bodies were small, more like the modern asteroids than planets or the moon. Just how small they were is a question that we must answer for each parent body, using evidence from the meteorites that sample it. In this chapter I shall consider some evidence that applies to all meteorites and then concentrate on that which bears specifically on the chondrites.

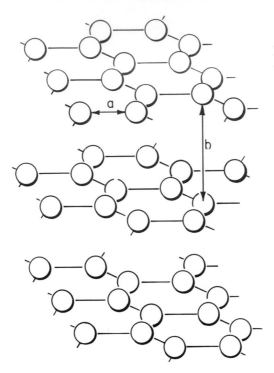

Figure 6.3 Crystal structure of graphite, greatly enlarged and with atoms separated for clarity, showing its short-strong (a) and long-weak bonds (b).

As a step toward determining how large a meteorite's parent body was, we can try to determine how deep it lay in that body. If we are lucky enough to choose a meteorite that formed at the center of its body—apparently the case for some kinds of irons—the depth we calculate will equal the body's radius. Otherwise, the depth will only give us a minimum body size. This is, unfortunately, the case for chondrites and many other kinds of meteorites.

Two measurable properties of a meteorite can give us an estimate of the depth of formation: the pressure at which its minerals formed and the rate at which it cooled from metamorphic or igneous temperatures. Both approaches take some explaining and neither works for all meteorites, but the results are interesting enough to make the effort worthwhile.

Polymorphism. Some chemical elements and compounds are natural thermometers or barometers: they take different crystal structures and forms under different conditions. Mineralogists call this ability of one chemical substance to exist in more than one form polymorphism. Carbon is an excellent example. At modest pressures, including those at and near the Earth's surface, it takes the form of graphite, in which the carbon atoms are arranged in stacks of honeycomb-like sheets (Figure 6.3). Because the carbon atoms in each sheet are held together very tightly, graphite can withstand very high temperatures and is therefore used in crucibles. The carbon sheets are bound together much more loosely and thus can slide past each other easily, making graphite one of the softest of minerals—hence its use in pencil "lead"—and a superb lubricant.

When graphite is subjected to immense pressures—variable with temperature, but roughly 60,000 times the Earth's atmospheric pressure (expressed as 60,000 atmospheres) at 1000° C—its atoms rearrange themselves into a more compact structure in which they are bound together very tightly in three dimensions (Figure 6.4). The mineral that results, diamond, is more dense than graphite and very much harder—harder in fact than any other mineral. At still higher pressures diamond gives way to a third polymorph of carbon, the rare mineral lonsdaleite.

Silicon dioxide, or silica, is far more versatile than carbon. Most familiar to us as the quartz that makes up most beach sand and is the chief ingredient of glass, silica occurs in eight other crystalline forms as well. Two of these, coesite and stishovite, form only at extremely high pressures; though they probably occur deep in the Earth as well, we have found them only at Meteor Crater and other meteorite impact sites. Two other silica polymorphs, cristobalite and tridymite, form at very high temperatures and are found only in igneous rocks. Tridymite is further restricted to volcanic and very shallow intrusive rocks, because it is unstable above a pressure that varies with temperature but never exceeds 3,000 atmospheres. Thus tridymite can be used as both a thermometer and a barometer.

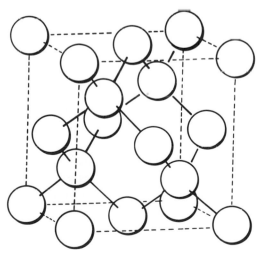

Figure 6.4 Crystal structure of diamond, showing the mineral's cubic symmetry and equally strong bonds in all directions.

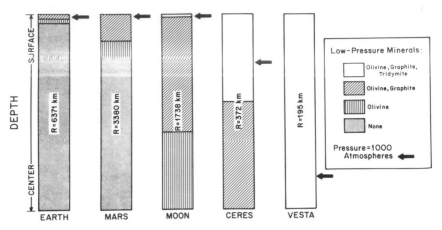

Figure 6.5 Relationship between depth and pressure in chondritic bodies with various radii. Patterns indicate which low-pressure polymorphs are present. Note that all such minerals (graphite, tridymite, olivine) persist throughout the large asteroid Vesta but are restricted to very shallow depths in the Earth.

The common iron-magnesium silicates olivine and the pyroxenes are less versatile than silica, but they too are transformed to different minerals at high pressures. For example, olivine gives way to the denser mineral ringwoodite at pressures greater than about 100,000 atmospheres, and pyroxene yields to majorite at still higher pressures. These high-pressure equivalents of olivine and pyroxene are of great interest to geophysicists concerned with the internal structure of the Earth, for their appearance seems to account for marked changes in the speed of seismic waves as they pass through the Earth's upper mantle.

Because these pressure-related transformations have been calibrated with laboratory experiments, we can use the presence or absence of high-pressure minerals to estimate the pressure at which a meteorite formed. Thus a meteorite that contains graphite but no diamond must have formed at a pressure less than about 60,000 atmospheres (at 1,000° C); one that contains olivine but no ringwoodite formed at a pressure less than about 100,000 atmospheres; and one that contains tridymite records a pressure no greater than 3,000 atmospheres.

As noted earlier, one argument raised in the early 1960s in favor of big parent bodies was that diamond occurs in one iron meteorite and several achondrites, specifically the ureilites. Figure 6.5 shows the basis for that argument; it indicates that graphite should persist to the center of any body that is much smaller than the moon.

The tally of high-pressure minerals in meteorites has grown considerably since 1960. Antarctica has yielded another diamond-bearing iron meteorite and several more ureilites, and both ringwoodite and majorite have been found in several ordinary chondrites, one of which is shown in Figure 6.6. However, our interpretation of these minerals has also changed. Every meteorite that contains high-pressure minerals also bears other features—fractures, deformed crystals—that testify to very intense impact shock. We now know that this process, rather than deep burial, was responsible for the extremely high pressures that produced diamond, ringwoodite, and majorite.

All *unshocked* meteorites known to us consist entirely of low-pressure minerals—graphite rather than diamond, olivine and pyroxene rather than ringwoodite and majorite. This fact places limits on the environments in which meteorites formed, but those limits are still lamentably broad. For example, the absence of diamond from meteorites that contain graphite tells us only that such meteorites did not form in the deep interiors of objects as big as the moon (see Figure

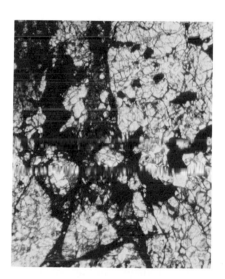 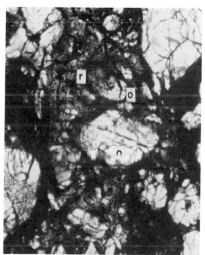

Figure 6.6 Photomicrographs, in transmitted light, of the Tenham, Australia, chondrite (L-group). (*left*) A shock-melt vein (dark) with abundant enclosed fragments of its host widens downward in the photograph, which is 2.0 mm (0.08 inch) wide. (*right*) Detailed view of the middle of the photograph shown. The large grain in the middle is olivine (o), which is surrounded by smoky purple ringwoodite (r; dark) of the same composition. The grain just above the large olivine crystal is part olivine (light), part ringwoodite. Dark glass lies between the crystals.

6.5). They could have formed at the center of a big asteroid like Ceres or at shallower depths in an object of any size.

We can place tighter limits on those few meteorites that contain free silica, in particular the enstatite chondrites and the closely similar enstatite achondrites. Since most of these meteorites contain tridymite, they could not have formed at pressures greater than 3,000 atmospheres, regardless of temperature. At the temperatures suggested by their mineralogy, the limit is still lower, between 700 and 1,000 atmospheres. As we can see from Figure 6.5, the enstatite meteorites could not have formed near the center of a body as big as Ceres. Once again, they could have formed at shallower depths in a larger body.

Solid solutions. Obviously, the presence or absence of high-pressure polymorphs gives us only a very crude idea of a meteorite's pressure history. For example, the absence of diamond from an ordinary chondrite that contains free carbon tells us only that the meteorite formed at some pressure less than about 60,000 atmospheres. To determine how much less, we need to find and measure some property that varies continuously with pressure. One such property is mineral composition, for although some minerals (for example, quartz) have just one composition, others (for example, pyroxenes) are *solid solutions:* they have compositions that vary regularly with temperature, pressure, or both.

To see why this is so, we need to know something about the structure of crystals. Linus Pauling, whose contributions to science and world peace have brought him two Nobel prizes, showed in 1925 that we can picture crystals as being made up of hard spheres—electrically neutral atoms or charged ions—that are packed together in efficient, geometrically regular arrangements and held together by balanced positive and negative charges. A well-packed crate of grapefruit (Figure 6.7, top section) shows such an arrangement, which very closely resembles the internal structure of a pure metal. An important implication of Pauling's model is that the size of an atom or ion strongly affects its ability to fit in a particular crystal structure. Pursuing the grapefruit analogy, it is obvious that we could not replace some of the grapefruit in Figure 6.7 with muskmelons without destroying the regular arrangement of the remaining grapefruit: a muskmelon is simply too big to fit in a hole designed for a grapefruit. We could replace some of the grapefruit with oranges, but the smaller fruits would then rattle around in holes that were too big for them. The crate would no longer be efficiently packed, in violation of Pauling's rules.

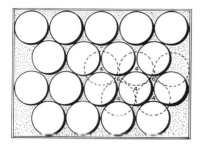

Figure 6.7 A closely packed crate of grapefruit, analogue for some crystal structures. (*top*) Overall view, showing one complete layer of fruit and, with dashes, part of a second layer. Numbers show spaces formed where four or six grapefruit come together. (*bottom*) Close-up view of 4-fold and 6-fold spaces or sites, which can accommodate smaller fruits or nuts.

If we wanted to put together a crate of mixed fruit without destroying the elegant symmetry of the grapefruit, we would have to use smaller fruits—limes, perhaps—and nuts. These would be too small to replace the grapefruit, but we could tuck them into the spaces between them. It turns out that the space formed where six grapefruit come together (Figure 6.7, bottom section) is just about big enough to accommodate a small lime, while the space formed where four come together is about right for a walnut. By separating the grapefruit slightly and filling some holes of each kind with limes and walnuts, we could produce a very elegant, tightly packed, and symmetrical crate of fruit and nuts.

This tightly packed crate provides a reasonable approximation of the structure of the mineral olivine. In this mineral, oxygen ions take the place of the grapefruit (Figure 6.8), silicon ions replace walnuts, and iron and magnesium replace limes. Of course, when we go from "neutral" fruits and nuts to charged ions, we have to make sure that the positive and negative charges balance. To do so, we fill enough holes so that there are two "limes" and one "walnut" for every four "grapefruit." This is the meaning of the formula of olivine that was presented in Table 3.2: X_2SiO_4, where X can be iron, magnesium, or both.

Note that two kinds of ions—iron and magnesium—have been substituted for the limes. The reason for this is evident in Figure 6.8, which shows the relative sizes of various common ions. Since doubly charged iron and magnesium ions have very similar radii, one substitutes freely for the other in olivine and many other silicates. This is true of manganese as well: most olivines contain small amounts of manganese, for the manganese ion is just a bit larger than iron.

Figure 6.8 shows that the calcium ion has about the same size relationship with oxygen as an orange has with a grapefruit. Just as oranges would not fit readily in the crate of mixed fruit—they are too big to replace limes or walnuts, too small to replace grapefruit—so calcium finds no suitable niche in the structure of olivine. We would expect olivine to be calcium-free, and most olivines very nearly are.

Thus olivine and other minerals have a limited number of environments or *sites* that can accommodate atoms or ions. It is for this reason that a particular mineral accepts some elements and rejects others. This fact has a great many implications; what is significant here is that these abilities and limitations change as physical conditions—in particular temperature and pressure—change. For example, mineralogists know that the taboo against calcium in olivine is relaxed in olivines that form at very high temperatures and cool rapidly, such as those in volcanic rocks. The reason why volcanic olivines contain small amounts of calcium is that heat expands the olivine structure, moving the oxygen ions apart and making room for calcium in the enlarged iron-magnesium sites. Rapid cooling then holds the calcium ions in sites that are really too small for them.

Pressure also affects the distribution of ions in minerals in predictable ways. For example, at high pressures, equal amounts of sodium and aluminum can replace iron and magnesium in pyroxenes. Since the extent of this substitution varies directly with pressure, we can measure the sodium and aluminum contents of a pyroxene and make a good estimate of the pressure under which, and hence the depth at

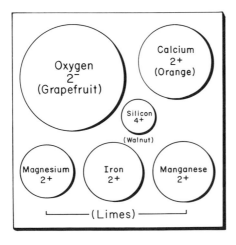

Figure 6.8 Relative sizes of familiar fruits and common ions. The fruits are smaller than life-size, the ions very much bigger: for example, the oxygen ion is really slightly more than 1/100,000,000th of an inch in diameter.

which, it formed. This approach has provided much of what we know about the mineralogy of the Earth's upper mantle, for lavas commonly bring up chunks of mantle rock. By using pyroxene compositions to estimate pressures, we can tell just how deep in the mantle these rocks came from.

Since pyroxenes are present in almost all kinds of meteorites, the same approach should work for these extraterrestrial rocks, and it does. A few years ago, one of my students showed that pyroxenes in metamorphosed LL-group ordinary chondrites contain so little sodium and aluminum that they must have formed at pressures no higher than 1,000 atmospheres—a big improvement on the "less than 60,000 atmospheres" indicated by the presence of carbon in the form of graphite. Though we know the compositions of other chondritic pyroxenes in less detail, we know that they contain very little aluminum. Thus they too formed at extremely modest pressures.

In summary, solid solutions take us a step further in our attempt to determine the pressures at which meteorites, in particular chondrites, formed. It appears that the ordinary and enstatite chondrites experienced pressures of less than about 1,000 terrestrial atmospheres. The arrows in Figure 6.5 show where this limiting pressure would occur in bodies of various sizes.

Cooling rates. A second way to estimate how deeply a meteorite was buried in its parent body is to determine how fast it cooled from metamorphic or igneous temperatures. The principle at work here is the same one involved in adding blankets to your bed on a cold night: the more insulating material—blankets or rock—that you place above a hot object, the slower it will cool. The pyroxenes and other minerals whose compositions vary with temperature can tell us how hot a meteorite was during metamorphism or crystallization, and we can make reasonable estimates of the insulating properties of rock. All we need in order to calculate the depth of burial is the cooling rate.

At present, there are two well-established ways to get this information. One makes use of the fact that the composition and structure of meteoritic metal change as a meteorite cools. The most obvious result of these changes is the Widmanstätten pattern observed in iron meteorites, which will be discussed in detail in Chapter 8. This approach works for most, though not quite all, meteorites that contain metallic nickel-iron.

A second, ingenious way to determine cooling rates makes use of plutonium-244, a very short-lived radioisotope. Though the solar system's original stock of this isotope decayed long ago, it was very

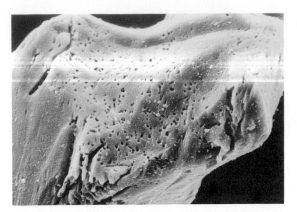

Figure 6.9 Plutonium-244 fission tracks in a crystal of pyroxene from the Tux Tuac (LL-group) ordinary chondrite. The crystal, about 80 micrometers wide, has been etched to reveal the tracks. (Photograph courtesy of Dr. Paul Pellas.)

much alive when the chondrites and most other meteorites formed. Because of its charge and ionic radius, nearly all of this isotope was concentrated in one of the minor minerals in meteorites: a phosphate. As plutonium-244 decayed in a meteorite, it yielded nuclear particles that shot through the phosphate and surrounding minerals to produce elongate damaged zones or *particle tracks*. These tracks, invisible in normal thin sections of meteorites, become enlarged when a sample is etched with a caustic solution and can thus be seen under a high-powered optical or electron microscope (Figure 6.9). By counting the plutonium-244 tracks in a measured area of a sample, we can determine the *track density* (number of tracks per unit area) in a phosphate grain or a mineral adjacent to it.

One way to put particle tracks to use is to employ the track density to determine how much plutonium-244 had decayed when the phosphate formed. This permits us to calculate the time between formation of the radioisotope and that of the meteorite. Geologists employ a similar approach, using tracks from long-lived uranium isotopes, to date some minerals that cannot be dated by other means.

To employ particle tracks to determine cooling rates, we make use of the fact that high temperatures destroy—anneal—tracks, and the temperature at which this occurs varies from one mineral to another. Thus as a meteorite cools, first one mineral "detector" and then another begins to retain its fission tracks. If a meteorite cooled very quickly (Figure 6.10a), its feldspar, pyroxene, and phosphate will have similar track densities, indicating that they began to record the decay of plutonium-244 at almost the same time. If it cooled slowly (Figure 6.10b), the track densities will decrease from feldspar to pyroxene to phosphate, indicating that the different minerals reached their retention temperatures at progressively later times. Since the retention temperatures for these minerals have been determined experimentally and we can calculate elapsed time from track densities, we have all the data we need to construct a time-temperature curve like the one shown in Figure 6.11.

Although it is easy to count particle tracks, preparation of samples for this sort of cooling rate determination is extremely laborious. Hence we have many cooling rates based on metal but rather few based on particle tracks. Until very recently, the shortage of the latter seemed to pose no problem, for cooling rates based on metal and on tracks agreed quite well. For example, both methods suggested cooling rates between 1° and 10° C per million years for most of the ordinary chondrites. However, new experimental work on nickel-iron alloys has thrown this agreement into question. This work, conducted by Joseph Goldstein and his students at Lehigh University, indicates that the cooling rates inferred previously for iron meteorites were too low by a factor of about 100. If this conclusion applies to the metal in chondrites as well, then there is a serious mismatch between the metallographic and fission-track cooling rates for these meteorites.

Obviously, uncertainty about metallographic cooling rates clouds our efforts to use these rates to infer the sizes of parent bodies. A few years ago, using the old cooling rate data for metal, I calculated that

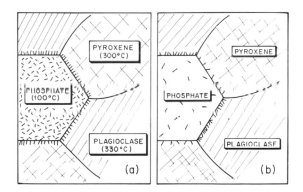

Figure 6.10 Particle track densities in a hypothetical phosphate crystal and adjacent plagioclase and calcium pyroxene, showing the influence of cooling rates. (*a*) With rapid cooling, all three minerals start to retain tracks at nearly the same time and have similar track densities. The temperature given for each mineral is that at which it retains 50 percent of the tracks that form. (*b*) With slow cooling, the three minerals start to record tracks in the sequence plagioclase–pyroxene–phosphate, and thus have different track densities.

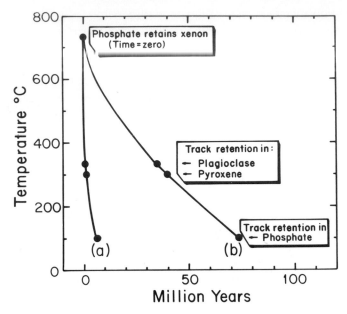

Figure 6.11 Time-temperature or cooling-rate curves inferred from retention of plutonium fission tracks in phosphate, pyroxene, and feldspar. Curve (a) corresponds to Figure 6.10a and suggests an average cooling rate of about 100 degrees per million years. Curve (b) corresponds to Figure 6.10b and suggests a rate of 8 to 9 degrees per million years.

the parent bodies of ordinary chondrites had minimum radii between 75 and 120 miles (Dodd, 1981). Scattered metal data for carbonaceous chondrites suggested somewhat larger bodies, perhaps 120 to 240 miles in radius. This size relationship is consistent with spectral reflectance data, which suggest that some very large asteroids—including the biggest, Ceres—are composed of carbonaceous chondritic material. Goldstein's new data on the cooling rates of iron meteorites, however, suggest that my estimates of the sizes of chondrite parent bodies may be too large by at least a factor of 10. Those bodies may have been just a few to a few tens of miles in diameter—very small indeed.

The Problem of Heat

This discussion of cooling rates and body sizes leads back to one of the main objections raised in the early 1960s against the formation of meteorites in small parent bodies: the problem of heating such bodies to temperatures high enough to cause metamorphism and, in some cases, melting. The source of heat, long a puzzle for meteoriticists, is made more so by the fact that chondrites and many other meteorites reached high temperatures within a few million years of their formation and cooled very swiftly thereafter. To explain these observations, we need a heat source that was present at the beginning of solar system history but vanished soon thereafter. The various candidates for this heat source are listed in Table 6.1.

Gravitational energy can heat a body at two stages in its history. As the body grows, its increasing gravitational pull draws material to it. When this material strikes its surface, much of its energy of motion (kinetic energy) is converted to heat. We find evidence of this at the bottoms of lunar and terrestrial meteorite craters, many of which contain masses of shock-melted rock. It is likely that intense bombardment by infalling material created the ocean of molten rock that covered the moon early in its history, and possible that a similar magma ocean covered the new Earth. However, this process requires that the incoming material be accelerated by a strong gravitational field. This in turn makes impact heating ineffective for bodies that are much smaller than the moon.

Gravitational energy is also released when molten metal trickles inward to form a planetary core, because the liquid is much denser than the silicates that it displaces. This mechanism can make planetary melting more or less self-sustaining, but it obviously cannot start the melting process. It is also more effective in large than in small bodies.

Gravitational energy may have heated planets early in their his-

Table 6.1 Energy sources for planetary heating

Energy source	Mechanism	Body size	Time
1. Gravity	a. Impact	Large	Early
	b. Core formation	Large	Early(?)
2. Long-lived radioactivity	Decay	Large	Late
3. Sun	a. Solar wind	Small	Early
	b. Luminosity	Small	Early
4. Tides	Friction	Any	Any
5. Extinct radioactivity	Decay	Any	Early

tories, but it was the slow decay of such long-lived radionuclides as uranium-235, uranium-238, and potassium-40 that kept them hot enough to produce magma for very long periods of time—4.5 billion years in the Earth's case, at least 1.5 billion years in the moon's. Such radionuclides produce heat very slowly. They can raise rock to melting temperatures in planetary bodies, where immense thicknesses of overlying rock act as insulation and hold the heat in, but they are far less effective in asteroidal bodies, which have much less insulation. Though the rocks that constitute such bodies probably produce nuclear heat as fast as those in planets, they lose it to space so rapidly that the interiors of asteroids are heated modestly, if at all.

This problem alone rules out long-lived radioactivity as the heat source for most meteorite parent bodies, but that source has another shortcoming as well. If a small, well-insulated body could manage to retain enough of the heat produced by long-lived radionuclides to reach metamorphic or igneous temperatures, it would reach them long after the body formed rather than at the beginning of its history. As we shall see in Chapter 9, a major reason behind the theory that a handful of achondrites sample a planet, probably Mars, is that they crystallized from molten rock billions of years after their parent body formed.

In 1960 it seemed to researchers that gravitational energy and long-lived radioactivity were the only sources of heat vigorous enough to account for high temperatures in meteorite parent bodies. It is therefore not surprising that most meteoriticists of the time regarded the problem of heat as a major stumbling block for small parent bodies.

All the other energy sources listed in Table 6.1 have been proposed or discovered since 1960. On the basis of astronomical observations of other stars at early stages in their history, we think the sun may have been much brighter 4.5 billion years ago than it is today, and its solar wind may have been far stronger. An extremely luminous sun would, of course, heat planetary surfaces directly. A powerful solar wind would heat them by electrical induction—a process not unlike microwave cooking. Both of these solar heating mechanisms meet two of the requirements for parent-body heat sources: they switched on during or shortly after the formation of the sun, and they switched off soon thereafter. Thus they are consistent with the early, brief high-temperature histories of chondrites and many other meteorites. They should also have been particularly effective for asteroidal bodies, for the low thermal and electrical conductivity of rocks makes it unlikely that they could heat the deep interiors of larger objects.

Unfortunately, we know too little about the histories of sun-like stars to be certain that the sun passed through a period of high luminosity or enhanced solar wind after the meteorite parent bodies formed. Indeed, the best evidence that it did would be proof that the meteorite parent bodies were heated from the outside. This is one of many instances in which the solution of a meteoritic puzzle would provide information of great value to astronomers.

The fourth entry in Table 6.1—tidal friction—came to the fore when images from the *Voyager* spacecraft showed us that Jupiter's innermost moon, tiny Io, is hurling out gouts of livid red and orange volcanic material. Since Io is much too small to be heated to melting temperatures by radioactive decay, the source of its heat is something of a mystery. The best suggestion to date is that tidal interaction with its gigantic neighbor is responsible: endless squeezing and relaxation of Io's rocks keeps them eternally hot. But Io's location close to Jupiter makes it a very special case. Though I have included tidal friction in Table 6.1 for the sake of completeness, it is obviously not a likely heat source for meteorite parent bodies.

The last entry in the table, extinct radioactivity, became a plausible heat source just a decade ago, when Typhoon Lee, Dimitri Papanastassiou, and Gerald Wasserburg of the California Institute of Technology showed that aluminum-26 was once present in the Allende chondrite. Though other short-lived radionuclides—for example, plutonium-244 and iodine-129—had been identified earlier through analysis of their decay products, none of them was abundant enough in the early solar system to be an important source of heat.

The approach that Lee and his co-workers used to detect the daughter of aluminum-26 in Allende is ingenious enough to be worth describing briefly. Because this isotope, magnesium-26, also forms in other ways, it is present in all magnesium-bearing minerals, for example olivine and the pyroxenes. Thus hunting for radiogenic magnesium-26 in such minerals is futile, rather like looking for a rare kind of quartz in a beach sand. Clearly, the Cal Tech team had to seek the daughter isotope in a mineral that contains abundant aluminum but no magnesium. Plagioclase feldspar meets this requirement, for its structure has no sites that accommodate the magnesium ion. Using a sensitive mass-spectrometer to analyze feldspar from Allende, Lee and his colleagues were able to detect magnesium-26 that could only have come from now-extinct aluminum-26.

Aluminum-26 was abundant in portions of the Allende meteorite—abundant enough to melt kilometer-sized bodies made of chondritic material with the same ratio of radioactive to stable aluminum. Unfortunately, we do not know how typical this ratio was, since it is far more difficult to identify radiogenic magnesium-26 in other kinds of meteorites. In spite of this reservation, most meteoriticists now accept extinct aluminum-26 as the most likely source of heat for the meteorite parent bodies: it was abundant when the solar system formed, and it decayed fast enough to account for the intense but brief thermal histories of most meteorites.

The Chondrite Parent Bodies

Meteoriticists no longer argue about whether chondrites come from asteroids or from much larger bodies, but we remain sharply divided on what their parent bodies were like—what we would see if we sliced a chondrite parent body like a melon and walked across it from side to side. Much of what follows is my own opinion, though I have tried to indicate where it is shared by others and where it is mine alone.

Ordinary chondrites. Meteoriticists know that each of the three groups of ordinary chondrites includes materials that range from virtually unmetamorphosed to strongly metamorphosed. We differ, however, on what these variations tell us about the interiors of the ordinary chondrite parent bodies.

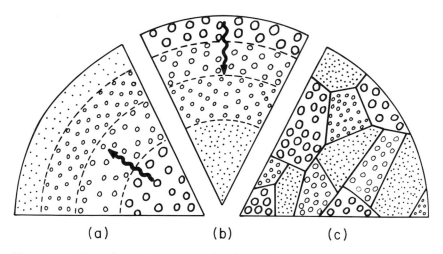

(a)   (b)   (c)

Figure 6.12 Hypothetical structures of ordinary chondrite parent bodies with various histories. Larger circles denote stronger metamorphism. (*a*) Body heated from within, for example by short-lived aluminum(Al)-26. (*b*) Body heated from without, for example by a super-luminous sun or enhanced solar wind. (*c*) Body composed of large chondritic fragments with various metamorphic histories inherited from a previous generation of bodies.

The conventional view, to which I subscribe, is that the progression from pristine to strongly metamorphosed material corresponds to increasing depth in the parent body (Figure 6.12a). According to this interpretation, each body was heated from within, probably by short-lived radioactivity. Unfortunately, there is little direct evidence to support this view. The strongest, at present pertinent only to the LL-group, is that the aluminum content of pyroxene, and hence pressure, increases with degree of metamorphism. Another interpretation is that metamorphic intensity increased *outward* as a result of external heating (Figure 6.12b). Some cooling rate studies appear to support this concept; others do not. A third interpretation, shown in Figure 6.12c, is that there is no regular relationship between metamorphic intensity and depth: that the chondrite parent bodies are huge breccias, made of a jumble of rocks with various previous metamorphic histories inherited from an earlier generation of tiny bodies. Though I find this complex picture hard to accept—it seems merely to push the problem back to earlier bodies—there is some evidence for it in the cooling histories of brecciated chondrites.

Each of these three models exists in two or more variations. Indeed, as one scientist said of the question of chondrule formation, there are almost as many interpretations as there are researchers considering the problem. Our view of the ordinary chondrite parent bodies stands today where the issue of body sizes stood a quarter-century ago: we have not yet found evidence that is strong enough to prove one interpretation and sweep all the others aside.

Enstatite meteorites. This is even more true for the enstatite chondrites and achondrites, with the additional complication that many workers believe that these meteorites formed in more than one parent body: the less metamorphosed chondrites in one, the more metamorphosed chondrites in a second, the achondrites perhaps in a third. This view is based on compositional and other discontinuities among the three kinds of meteorites.

Though these discontinuities are real enough, I am more impressed by the fact that all three kinds of enstatite meteorites share a common chemical trend that appears, on present evidence, to reflect partial melting. My view, very much in the minority today, is that the enstatite meteorites formed in one asteroid, whose internal structure resembles that sketched in Figure 6.13. According to this interpretation, chemical differences among the chondrites and achondrites are due to melting of metal and sulfide. one result of which was the formation of a nickel-iron core. Samples of that core may survive in one group of iron meteorites (see Chapter 8).

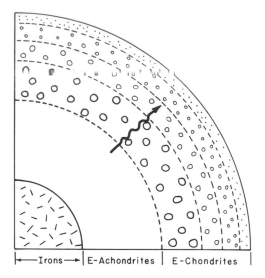

Figure 6.13 Hypothetical cross-section of a parent body heated from within and composed of enstatite chondrites, enstatite achondrites, and iron meteorites. According to this interpretation, the irons represent metallic liquid derived from E chondrites, the E achondrites the solids that remained after melting.

If the drawing in Figure 6.13 is correct, then enstatite meteorites bridge the gap between primitive and differentiated meteorites. Since many of the enstatite meteorites are complex breccias, it will be hard to verify both the model presented here and alternatives to it. Nonetheless, the possibility that we have sampled one asteroid from surface to center makes the enstatite chondrites and achondrites among the most intriguing of meteorites.

Carbonaceous chondrites. The carbonaceous chondrites, very few of which experienced metamorphism, stand aside from arguments about whether parent bodies were heated from within or from without, but their very freedom from metamorphism raises other problems. The strongest evidence for the source of heat for this process comes from a carbonaceous chondrite, Allende, and it suggests that this source—aluminum-26—was potent enough to melt very small asteroids. How can we reconcile this fact with evidence from cooling rates and reflectance spectra that the carbonaceous chondrites in fact resided in relatively large bodies?

The answer may lie in two other properties of carbonaceous chondrites: first, many of them show evidence of chemical alteration by water; second, their radiometric ages suggest that such alteration continued long after the meteorites formed. Perhaps the parent bodies of carbonaceous chondrites consisted of rocky material mixed with large amounts of ice—something like the mix that makes up comet nuclei. Heat generated by aluminum-26 at first went to melt the ice, providing the water that altered silicates in these meteorites. As the ice was dissipated, the rocky material that remained was brought together to form the meteorites as we see them today. According to this interpretation, the carbonaceous chondrites accreted twice: once when their asteroidal parent bodies formed as masses of rock and ice, again when the ice melted. In this case my unorthodox interpretation has little competition; we have just begun to think about the structure and history of carbonaceous chondrite parent bodies.

With this review of what we know—and still do not know—about the parent bodies of chondrites, we shall lay these primitive meteorites aside and turn to the differentiated meteorites, whose igneous origin is beyond question. We shall return to the chondrites in Chapter 10, where we'll see what meteorites can tell us about the first stages in the history of the solar system.

7

When
Planets Melt

Meteoriticists are particularly interested in chondrites because they remember early stages in their history that all other rocks have forgotten. Nearly all of them, with the possible exception of some enstatite chondrites, chemically resemble the primitive nebular material from which they formed 4.5 billion years ago. A precious few chondrites, those that escaped thermal metamorphism in their parent bodies, resemble that material physically as well, giving us glimpses of the processes through which a swirl of dust and gas evolved into a star and its retinue of planets, satellites, asteroids, and comets.

The differentiated meteorites—irons, stony-irons, and achondrites—have much shorter memories, for the melting and crystallization that gave them their nonsolar chemical compositions also erased textural and mineralogical evidence of their earlier history. Though these meteorites can give us insights into processes of planetary evolution, they tell us little about how planets formed and still less about events in the primitive solar nebula.

The historical differences between chondrites and differentiated meteorites compel us to study them in different ways. In this book I have approached the chondrites by describing them and working backward from their properties to the processes that are responsible for them. I shall take the opposite tack with differentiated meteorites, first discussing how differentiation takes place in more familiar materials and then, in the next two chapters, applying that knowledge to various kinds of differentiated meteorites. This strategy is appropriate for these igneous meteorites, for although no one has ever seen a

94

chondrite form and no one ever will, our long experience with the hot, active Earth has taught us a great deal about igneous rocks.

Differentiation

We can define differentiation as the separation of one kind of material into two or more kinds with different chemical compositions. Familiar examples are the separation of cream from whole milk, crystallization of salts on a kitchen faucet, and distillation of sour mash to produce bourbon. In the first case, one liquid (or more accurately, an intimate mixture or emulsion) divides into two liquids; in the second, solids crystallize out of a liquid; in the third, an alcohol-rich vapor bubbles out of the mash. The common factor in all three examples is that the products are physically distinct and chemically different from the starting material.

Although each of these three types of differentiation—liquid-liquid, liquid-solid, liquid-vapor—has played a role in the history of igneous rocks and meteorites, the most important type is that which involves separation of crystals from a liquid, through melting or crystallization. Thus we can best begin our consideration of differentiated meteorites with a brief discussion of the consequences of these processes. I shall start with some simple and familiar examples of melting and crystallization and progress from these to igneous rocks and, ultimately, planets.

It is common knowledge that water freezes at 32° F (0° C). If we put a tray of water in the freezer and wait, we find after an hour or two that the tray is full of ice. If curiosity leads us to put a thermometer in the water and check its temperature from time to time, we will find that the water cools to 32 degrees, holds steady at that temperature while freezing takes place, and then resumes cooling after all of the water becomes ice. Clearly, pure water freezes in a very simple way. We can lower its freezing point by applying pressure, a tactic that ice skaters use without thinking about it, but at a given pressure—for example that of the Earth's atmosphere—water freezes at just one temperature to form a solid of the same composition.

This refrigerator experiment is simple but not very interesting. Aside from ice itself—a rock in very cold parts of the world, though seldom regarded as such—few rocks consist of just one chemical substance. We can make the experiment much more instructive and only a bit more complex by adding table salt (sodium chloride) to the water and taking it outside.

If you live in the northern part of our country, you have probably scattered salt on an icy sidewalk and observed that it "melts the ice." (What it actually does is combine with the ice to form a mixture with a lower freezing temperature; a home ice cream machine uses the same principle.) You may have noticed too that it takes more salt to do the job on a very cold day than it does when the temperature is just below freezing. Within limits, the more salt you mix with the ice, the lower is its melting temperature. Obviously salt water, unlike the pure water in your ice cube tray, freezes at various temperatures depending on how much salt is dissolved in it.

If you live in West Yellowstone, International Falls, or some other place that regularly experiences record low temperatures, you know from hard experience that salting an icy sidewalk does no good if the outside temperature is exceptionally low: no amount of salt will clear the walk if the temperature is very far below freezing. If you have tried to melt ice under such conditions, you know that the salt just lies there until the weather warms up a bit. What you may not know is that when melting finally begins, it always does so at the same temperature—about $-6°$ F—and that, regardless of the amount of salt that has been lavished on the sidewalk, the first liquid that forms always contains the same proportions of salt and water, about 23 percent of the former by weight.

These observations on crystallization and melting can be summarized in a simple diagram. Figure 7.1 shows how the freezing temperature of water changes as salt is added. Above the curved line, only liquid is possible; we therefore call that line the *liquidus*. Below the straight horizontal line, only solids are possible; hence we call that line the *solidus*. Between the lines, one solid—either ice or salt—coexists with salt water. The figure also shows that, at a pressure of one atmosphere, pure water freezes at 32° F but water that contains up to 23 percent salt starts to freeze at lower temperatures. At higher concentrations of salt, the freezing temperature rises again—a reflection of the observation that beyond a certain point, salting an icy sidewalk is futile. That point—the concentration of salt for which the freezing temperature is lowest—is called a minimum melting point or *eutectic* (point c).

What happens when we cool salt water, say a solution that contains 10 percent salt (composition a in Figure 7.1)? Nothing at all happens until its temperature reaches the liquidus. At that point, a tiny amount of ice forms and the remaining water becomes slightly saltier. But we know that this saltier water freezes at a lower temperature, and thus for more ice to form, the temperature must continue to fall. As it does so, we produce more and more ice and leave less and less

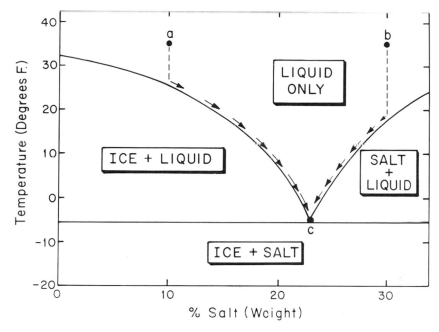

Figure 7.1 Melting and crystallization of ice-salt mixtures at the pressure of the Earth's atmosphere.

water; at the same time, the water that remains becomes saltier. This process continues until the liquid composition reaches the eutectic. At that point, the temperature stops falling and stays constant while the remaining liquid crystallizes to form a mixture of 23 percent salt and 77 percent ice.

There is nothing sacred about composition a. If we repeated this experiment with other liquid compositions between those of pure water and the eutectic, we would observe the same sequence of events: crystallization of ice, followed, at the eutectic, by crystallization of an intimate mixture of ice and water. The only difference from one starting composition to another would be in the amount of ice that formed before the liquid reached the eutectic: large for starting compositions close to pure water, small for compositions close to the eutectic.

What kinds of "rocks" can we produce by cooling liquid a? This depends on how we conduct the experiment, in particular on whether we leave the crystals that form in the pot or remove them. If we let the experiment run undisturbed, the product will look like the drawing in Figure 7.2a: a few large ice crystals set in a matrix of tiny, intimately intergrown crystals of salt and ice. The large ice crystals represent crystallization of that substance alone, the intergrowth simultaneous crystallization of salt and ice at the eutectic. This "rock" looks very much like what geologists call *porphyry:* an igneous rock composed of large, early-formed crystals set in a matrix of small, late-formed crystals. Since we have neither added nor removed material to form this porphyritic "rock," its chemical composition is that of the original liquid: 10 percent sodium chloride, 90 percent water. In other words, our experiment has not resulted in chemical differentiation.

Parts b and c of Figure 7.2 show two "rocks" that would result if we removed ice crystals from the experiment as they formed, for example by letting them float to the surface. The "rock" in Figure 7.2b consists entirely of large ice crystals and is thus made up of pure water. The one in Figure 7.2c, made up wholly of material that crystallized at the eutectic, is texturally similar to the matrix in Figure

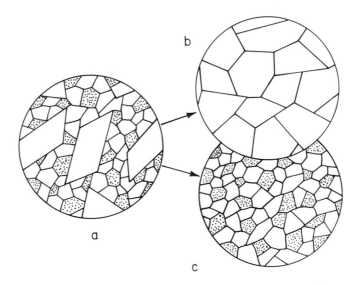

Figure 7.2 "Rocks" formed by crystallizing a solution of 10 percent salt in water (point a in Figure 7.1). Clear crystals are water ice, stippled ones are salt. (*a*) Total crystallization yields a rock composed of big, well-formed ice crystals in a finer-grained ice-salt mixture. The rock has the same chemical composition as the original liquid (*porphyritic* texture). (*b*) Accumulation of early-formed ice crystals yields a coarse-grained rock composed entirely of water (*cumulate* texture). (*c*) The liquid that remains when ice is removed crystallizes to a fine-grained ice-salt mixture with the eutectic composition.

7.2a. Like that material, it consists of intergrown salt and ice crystals with a total composition of 23 percent sodium chloride and 77 percent water.

It should be obvious that separation of crystals from liquid, what geologists call *fractional crystallization,* can yield a wide variety of products from one starting composition. By varying the conditions of the experiment—for example, removing only a part of the ice that forms—we can produce rocks with varied textures and any composition between pure water and 23 percent sodium chloride. However, there is a limit to our ability to tamper with the experiment: no liquid that starts at or to the left of the eutectic in Figure 7.1 can produce large crystals of salt or rocks that contain more than 23 percent sodium chloride by weight. Liquids to the *right* of the eutectic can do so, for crystallization of such liquids (for example, b) mimics that of liquid a, but with salt crystals forming first.

In summary, when solutions of salt and water crystallize, they do so over a long range of temperature, with either ice or salt crystallizing first and the composition of the remaining liquid traveling down the liquidus to the eutectic point. Only pure water, pure molten sodium chloride, and a solution of eutectic composition—23 percent sodium chloride—crystallize at one temperature to form a "rock" of the same composition. For all other starting compositions, fractional crystallization can yield "rocks" with widely varied textures and compositions.

Basalt: A Planetary Leitmotif

Figure 7.1 carries one more message that is important for what follows. Just as the eutectic point represents the last drop of liquid that remains when solutions freeze, so it represents the first drop that forms when solid mixtures of ice and salt start to melt. Regardless of the proportions of salt and ice in such mixtures, they start to melt at $-6°$ F (at a pressure of one atmosphere), and the first liquid that forms contains 23 percent salt.

We might expect mixtures of many more elements to melt in a more complex fashion, and they do, but they too tend to produce early liquids with similar chemical compositions. For a wide range of more or less chondritic starting materials, these early liquids crystallize to form rocks composed largely of pyroxenes and plagioclase. The most common of these rocks on Earth are called *basalts,* and the magmas from which they form are called *basaltic.* Basalts are the most widespread of earthly rocks. They cover the ocean bottoms, from which

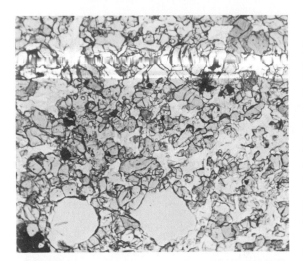

Figure 7.3 Basaltic meteorites, viewed in thin section with transmitted light. (*top*) The Ibitira eucrite, composed chiefly of pyroxene (dark) and plagioclase (light). Round vesicles reflect bubbles in the magma and suggest that this meteorite samples a lava flow. The field of view is 1/16 inch wide. (From Dodd, 1981.) (*bottom*) The Sherghotty achondrite resembles coarse-grained eucrites in mineralogy and texture, but its pyroxene (gray) shows strong shock deformation and its plagioclase (white) has been converted to a shock glass, maskelynite. Field of view is 1/8 inch wide. (Photomicrograph courtesy of Professor H. Y. McSween.)

they rise in huge volcanic edifices like the Hawaiian islands and Iceland. They also appear on the continents, where they sprawl over vast areas such as the Columbia Plateau of the northwestern United States, the Karroo of South Africa, and the Deccan Plateau of India. Though rocks from these different areas are by no means chemically identical—their magmas came from different depths within the Earth, and their compositions vary somewhat with both the pressure at the source and source composition—they are similar enough to justify calling all of them basalts.

Long before Apollo 11's *Eagle* touched down at Tranquillity Base, geologists suspected that basaltic volcanic rocks also cover the moon's dark lowlands or maria. Samples returned by the Apollo and

Luna missions confirmed that suspicion, and unmanned probes to Mars and Venus have shown that similar rocks are abundant on those planets as well. It is clear that basaltic volcanism is a repeated theme, or leitmotif, of planetary evolution.

This leitmotif was also played on at least two of the meteorite parent bodies. Figure 7.3 shows photomicrographs of two basalt-like achondrites, a eucrite and a sherghottite, which crystallized from basaltic magmas under somewhat different conditions. The eucrite, texturally similar to some terrestrial and lunar basalts, is a fine-grained mixture of calcium-rich pyroxene and plagioclase that formed on the surface of its parent body; like some terrestrial basalts, it reflects rapid crystallization with very little, if any, crystal-liquid differentiation. Some eucrites and the sherghottite shown in the bottom photograph are coarser-grained and reflect slower crystallization at greater depths in their parent bodies.

The two rocks in Figure 7.3 are so similar in mineralogy and chemical composition that earlier researchers lumped eucrites and sherghottites together under the term *basaltic achondrites* and inferred that they sample the same parent body. As we shall see in Chapter 9, chemical and isotopic differences between eucrites and sherghottites and their very different times of formation indicate that these two groups of igneous achondrites in fact come from different parent bodies.

Layered Intrusions

On the Earth and on all other bodies that are big enough to have significant gravity fields, density differences between crystals and liquid expedite differentiation. Molten rock or magma is less dense than its solid equivalent and tends to rise toward the surface. Likewise, some crystals (for example, olivine and the pyroxenes) settle through magma and others (plagioclase) float in it. As is true in the simple example of salt and water, these processes of crystal-liquid fractionation can yield rocks of varied mineralogy and texture whose chemical compositions differ dramatically from the composition of the original magma.

Crystallization of basaltic magmas is much more complex than that of salt water, but it is worth considering briefly because it bears on the origin of many kinds of differentiated meteorites. Long before geologists learned how to simulate crystallization of magmas in the laboratory, they knew that large bodies of basaltic magma that crystallized slowly within the Earth's crust have a characteristic layered structure,

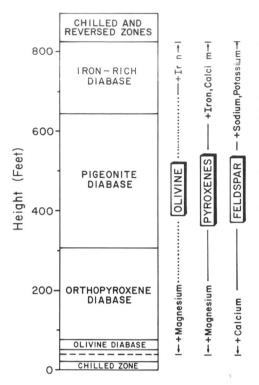

Figure 7.4 Diagrammatic cross section of the Palisades sill of New York and New Jersey, showing variation of its mineralogy with height. The intrusion consists of diabase, a coarser-grained equivalent of basalt. Olivine appears near the bottom (iron-poor) and top (iron-rich), but is absent elsewhere. The dashed line at the bottom of the diagram marks the boundary between two masses of magma that formed the intrusion.

with some minerals concentrated near the bottom, others near the top, still others in between.

One such *layered intrusion* makes up the Palisades, a brow of dark gray rock that forms the western bank of the Hudson River opposite New York City. A cross-section of this 900–1,000 foot sheet, or *sill,* is shown in Figure 7.4. The Palisades sill formed about 185 million years ago, when two batches of basaltic magma slipped between layers of the surrounding sedimentary rocks. The first intrusion, a very modest one, is represented by rocks below the dashed line in Figure 7.4. It froze so quickly against the cold rocks above and below it that rocks at the contact—the chill zone—are very fine-grained. The second Palisades intrusion, shown above the dashed line in Figure 7.4, was much thicker, and the rocks it produced record a more interesting history. Though most of them are *diabase,* coarser-grained but chemically equivalent to basalt, their mineralogy varies regularly with height in the sill. Near the bottom of the second intrusion is a layer enriched in magnesian olivine—the "olivine diabase" in Figure 7.4. Above that, various pyroxenes appear in an orderly sequence, from magnesium-rich, calcium-poor varieties (orthopyroxene, pigeonite) near the bottom to iron-rich and calcium-rich varieties (augite) near the top. These pyroxenes are accompanied by plagioclase feldspar, which is present throughout the sill but becomes more abundant and richer in sodium in the upper layers.

The chemical composition of the Palisades diabase changes regularly with its mineralogy. Olivine diabase near the bottom of the sill contains much more magnesium and less silicon than the original magma. Rocks near the top of the sill are, in contrast, depleted in magnesium and enriched in iron, aluminum, silicon, potassium, and sodium. Indeed, some rocks in the zone labeled "iron-rich diabase" in Figure 7.4 approach the chemical composition of granite, a rock composed almost entirely of sodium-potassium feldspars and quartz.

The Palisades sill is a frozen experiment in fractional crystallization driven by gravity. Olivine crystallized early and settled toward the bottom of the sheet to form the olivine diabase layer. Iron-magnesium and iron-magnesium-calcium pyroxenes followed, as did calcium-rich plagioclase. As these minerals settled out, the remaining liquid changed composition accordingly. A thin zone at the top of the sill records the same sequence of mineralogical and chemical changes on a smaller scale and in reverse order. Norman Bowen, one of this century's most accomplished students of igneous rocks, showed that the pattern of crystallization recorded in the Palisades is repeated, with some variations, in other layered intrusions. The usual sequence of crystallization, which geologists call Bowen's "reaction series," is outlined in Figure 7.5.

The efficiency of fractional crystallization in layered terrestrial intrusions depends on the speed at which they cooled. Because the Palisades intrusion is thin and it was emplaced near the Earth's surface, it cooled rapidly. Hence its layering is subtle: only its olivine diabase layer shows conspicuous enrichment in one mineral. Layering is much more pronounced in very thick intrusions such as South Africa's Bushveld Complex and Montana's Stillwater Complex. Figure 7.6 shows a cross-section of the lower part of the latter intrusion, only 60 percent of which is exposed.

The Stillwater Complex is layered at every scale, from inches to thousands of feet. Most of it consists of coarse-grained basaltic rocks (called *gabbro*), but layers composed of just one or two minerals are very common. Thus the thick "ultrabasic zone" at the base of the complex includes individual layers that consist almost entirely of either olivine, pyroxene, or the iron-chromium oxide *chromite*. Higher in the complex we find layers of *anorthosite,* a rock composed principally of plagioclase feldspar. These one-mineral rocks testify to

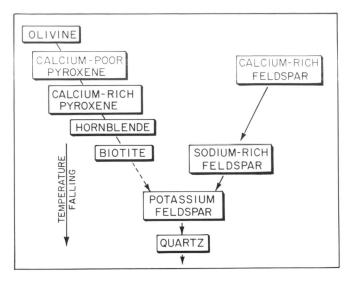

Figure 7.5 Crystallization sequence for basaltic magmas, as outlined by N. L. Bowen. Hornblende and biotite are water-bearing silicates, the former very rare in and the latter absent from meteorites.

the high efficiency of crystal settling, or, in the case of plagioclase, flotation. (They also explain why such huge layered complexes are our principal source of chromium, an important component of steel.)

Accumulation of crystals is the most common way by which igneous rocks come to consist almost wholly of one or two minerals, and we therefore call such rocks *cumulates*. Cumulate rocks, a hallmark of layered intrusions on Earth, are also remarkably common among differentiated meteorites (Table 7.1). For example, the Chassigny meteorite (Figure 7.7, top) consists almost entirely of olivine and looks strikingly like the terrestrial olivine rock dunite. The pallasites (see Figure 3.3) are also olivine cumulates, albeit with metal rather than other silicates between the large olivine crystals. Other differentiated meteorites consist almost entirely of one or more pyroxenes, the diogenites of the iron-magnesium pyroxene bronzite, the enstatite achondrites of almost pure magnesium pyroxene, and the nakhlites of the calcium-iron-magnesium pyroxene *augite* (Figure 7.7, bottom). Most of these meteorites, like their earthly analogues, are products of crystal settling (see Table 7.1).

One more important point should be made about the chemical consequences of fractional crystallization. As crystals form and settle (or float), they take some elements out of the magma and leave it relatively enriched in others. Thus olivine in terrestrial layered intrusions removes iron, magnesium, manganese, and other ions—such as nickel—that fit comfortably in the structure of that mineral, leaving the liquid enriched in, for example, calcium, aluminum, and sodium. Plagioclase accommodates these elements, plus other minor elements—such as strontium—whose ions can substitute for these constituents in the plagioclase structure.

In the course of fractional crystallization, all of the more abundant elements in a magma and many of the minor ones find homes in one mineral or another, but the residual liquid accumulates a wide variety of elements—copper and sulfur, for example—whose ions fit in none of the igneous minerals. These elements, so sparse in the original magma that they are hard to detect chemically and impossible to recover commercially, may eventually reach useful concentrations. Thus, for example, veins that formed during the last stages of crystallization of the Palisades diabase are rich in copper sulfides, and these veins were once the basis of a bustling copper industry in New Jersey.

New Jersey's now-defunct copper industry lies far from the subject of this book, but the fate of trace elements in magmas does not. As we shall see in Chapters 8 and 9, these elements can tell us a great deal

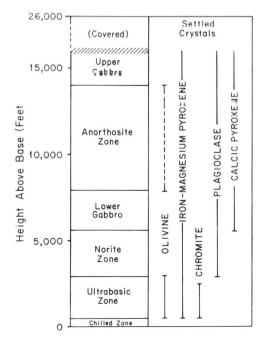

Figure 7.6 Cross-section of the exposed lower portion of the Stillwater Complex, Montana. Most rocks of the complex are gabbros, coarse-grained equivalents of basalts. Norite is a gabbro that is enriched in calcium-poor pyroxene, and anorthosite a rock composed largely of plagioclase feldspar. Vertical bars show the distribution of settled (cumulate) crystals. Each mineral is concentrated locally in layers that are nearly free of other minerals.

Table 7.1 Physical histories of differentiated meteorites

History	Major constituents	Meteorites
1. Partial melts with		
a. little fractionation	Pyroxenes, feldspar	Eucrites
b. much fractionation	Nickel-iron, sulfide	Irons
2. Crystal cumulates	Olivine	Chassignite
	Olivine + Ca-poor pyroxenes	Urcilitcs
	Olivine + Ca-poor pyroxenes + metal	Lodranites
	Olivine + metal	Pallasites
	Ca-, Fe-poor pyroxenes	Enstatite achondrites
	Ca-poor Fe-Mg pyroxenes	Diogenites
	Ca-rich pyroxene	Angrite
	Ca-rich pyroxene + olivine	Nakhlites
	Ca-rich pyroxene + plagioclase	Eucrites, sherghottites
3. Breccias	Eucrites + diogenites	Howardites
	Eucrites + diogenites + metal	Mesosiderites

Note: Some meteorite groups include materials with somewhat different histories. For example, some members of nearly all groups are breccias.

about the chemical makeup and history of bodies that have yielded just a handful of meteorites.

Layered Planets

By building stepwise from the crystallization of very simple mixtures to the crystallization of magmas in layered intrusions, we have seen that many differentiated meteorites resemble the products of fractional crystallization (see Table 7.1). So far, the examples used have been either well known to everyone (ice/salt) or at least well understood by geologists. The next leap, to layered planets, carries us from the firm ground of observation and experiment to a bog of speculation. For though we know that the Earth and moon are layered bodies, and we suspect that this is true of the other terrestrial planets as well, it is by no means clear how this layering came about. Did the Earth's core and mantle and its primitive crust develop just after our planet formed, or did they evolve gradually through repeated episodes of partial melting? Was the Earth once a homogeneous ball of chondritic material, or was its deep interior always metal-rich?

Fortunately, these issues concern geologists more than meteoriticists. For the present discussion, we shall assume that at least some bodies in the solar system—those that housed differentiated meteorites—began as homogeneous balls of chondritic material and ask what happened when those bodies started to melt. We learned from the discussion of salt and water that chemical mixtures do not melt directly to liquids of the same composition; rather, a liquid of low melting temperature appears first. If that liquid is much more (or less) dense than the remaining solids, it may sink (or rise) to form a separate pool of differentiated material.

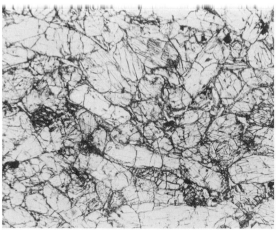

Figure 7.7 Photomicrographs of cumulate achondritic meteorites. (From Dodd, 1981.) (*top*) Chassigny, shown in doubly polarized light in a view about 1/16 inch wide. The meteorite consists almost entirely of olivine (white to gray), with small amounts of pyroxene and chromite (black). (*bottom*) Nakhla, Egypt, one of three known nakhlites, shown in transmitted light in a field about 1/16 inch wide. Most large crystals, arranged in a rough parallel alignment, are the calcium-rich pyroxene augite.

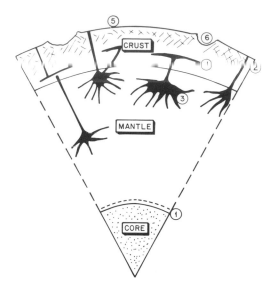

Figure 7.8 A hypothetical differentiated body of L-chondrite composition, showing possible environments of formation for various differentiated meteorites. Irons may sample the core. Other sources and products are as follows: 1 = core-mantle boundary (pallasites); 2 = impact-brecciated crust (howardites, mesosiderites); 3 = magma sources; 4 = shallow intrusions (coarse eucrites and sherghottites; cumulate achondrites); 5 = surface flows (fine-grained eucrites, sherghottites); 6 = impact craters, responsible for brecciation. (Note that this diagram summarizes the experiences of meteorites from several different parent bodies.)

Chondritic meteorites also melt in stages. At first they yield a liquid composed chiefly of metallic iron and nickel, sulfur, and other elements that tend to combine with either iron or sulfur. Because such a liquid is more than twice as dense as chondritic silicates, it moves downward under the influence of gravity. If it accumulates in the center of the body, the result is a dense mass of metal and sulfide: a *core*. Further melting of chondritic material yields liquids of basalt-like composition and leaves behind solid residues composed of minerals—chiefly olivine and pyroxenes—that melt at still higher temperatures. Since the basaltic liquids are *less* dense than the remaining solids, they tend to rise. Some may reach the surface and flow forth as lava; others may stop below the surface and crystallize slowly to form layered intrusions like the Palisades and the Stillwater Complex. Whatever their detailed history may be, the upward migration and crystallization of basalt-like liquids produce a *crust* composed largely of pyroxenes and feldspar. These processes leave behind a solid residue—a *mantle*—composed largely of olivine. Thus, the melting of a large ball of chondritic material may produce a three-layered body rather like our layered Earth. Such a body is shown in Figure 7.8.

In preparing Figure 7.8, I assumed that the parent body has the bulk composition of L-group ordinary chondrites and that melting did not change the distribution of iron between metal and silicates. Thus rocks of the crust and mantle contain the same proportions of magnesium and oxidized iron that we find in L chondrites, and the core has the iron/nickel ratio of L-chondrite metal. I also assumed that melting produced a crust composed almost wholly of pyroxenes and plagioclase and left a mantle composed of olivine.

Since chondrites have varied compositions, we might expect a parent body derived from carbonaceous or enstatite chondrites to be somewhat different from the one shown in Figure 7.8. As Figure 7.9 shows, even bodies composed of the same material—once again L chondrites—can evolve in different ways according to the availability

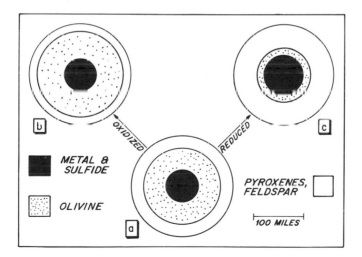

Figure 7.9 Three hypothetical parent bodies of L chondrite composition with different histories of oxidation or reduction. (*a*) Iron is distributed among metal, sulfide, and silicates as it is in L chondrites. (*b*) Most iron is oxidized and present in olivine and pyroxenes. (*c*) Most iron is reduced and present in metal and sulfide.

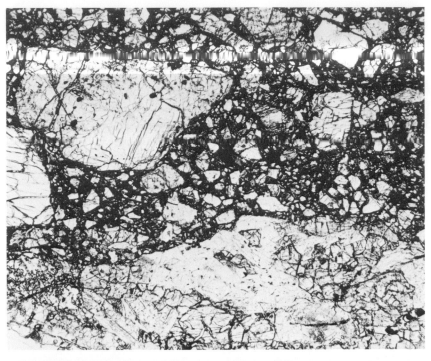

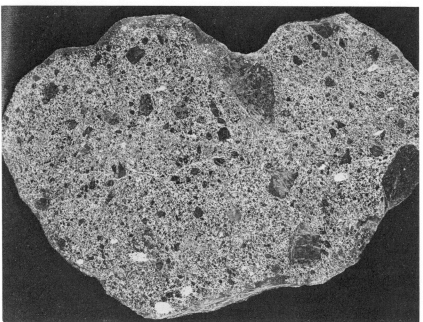

of oxygen. In constructing Figure 7.9a, I made the same assumptions that I used for Figure 7.8: that the parent material neither gained nor lost oxygen during melting—that is, it was neither oxidized nor reduced.

In constructing body b, I assumed that oxygen was readily available during melting—that it was abundant enough to convert most of the iron to the doubly charged ion that fits in olivine and pyroxene. Thus body b's small core consists of iron sulfide and metallic nickel. Its mantle and crust contain silicates that are richer in iron than those of body a. Oxidation is also responsible for body b's very thick mantle. As we saw in Table 3.2, even though both olivine and the pyroxenes contain iron, magnesium, and silicon, the ratio of the first two elements to the third is different in the two kinds of minerals: two in the former, one in the latter. Thus increasing the abundance of oxidized iron in silicates shifts the olivine:pyroxene ratio toward olivine.

When I constructed body c, I made just the opposite assumption: that oxygen was in very short supply, with the result that all of the iron in that body was converted to sulfide or metal. Thus metal in the core of body c has a very high iron:nickel ratio, while the olivine and pyroxenes in this body's mantle and crust are iron-free. In this body too, tampering with the distribution of iron affects the proportions of olivine and pyroxene and hence the thicknesses of the mantle and crust. In this case, removing iron from the silicates *lowers* the ratio of iron plus magnesium to silicon to produce pyroxenes at the expense of olivine. The resulting increase in pyroxenes is expressed in Figure 7.9c as a very thick crust.

As we shall see in Chapters 8 and 9, reduction and oxidation played important roles in the evolution of differentiated meteorites. For the moment, however, the most important point to be made about Figure 7.8 is that we can explain most kinds of differentiated

Figure 7.10 (*opposite*) Brecciated differentiated meteorites. (*top*) Thin section of the Jodzie howardite, a surface breccia. The larger fragments include a eucritic clast (bottom) and several fragments of iron-magnesium pyroxene. Similar materials make up the matrix, which is so fine-grained that it appears opaque in this photograph. The field of view is 1/8 inch wide, and the sample is viewed in transmitted light. (From Dodd, 1981.) (*bottom*) Polished slab of the Crab Orchard Mountains mesosiderite, a surface breccia composed of achondritic clasts (dark) and metal (light) in similar proportions. The slab is about 9 inches wide. (Photograph by Thane Bierwert, courtesy of the Department of Library Services, American Museum of Natural History; negative no. 314903.)

meteorites in terms of melting and fractional crystallization within their parent bodies. Clearly, the iron meteorites may sample its core and the pallasites its core-mantle boundary. The basalt-like eucrites and sherghottites represent more or less undifferentiated silicate melts, while the various cumulate achondrites are products of fractional crystallization of such melts in the crust.

Although most differentiated meteorites are igneous rocks, some—the howardites and mesosiderites in particular—are breccias composed of fragments of such rocks (Figure 7.10). To account for these meteorites in Figure 7.8, we must assume that impacts shattered and mixed rocks on the surface of our parent body, as they did on the moon.

Figure 7.8 looks suspiciously like the single, large meteorite parent body that Daly envisioned in 1943. Unfortunately, when we look at the differentiated meteorites in detail, as we shall in the next two chapters, we find compelling evidence that they really sample many different bodies: some come from the core of one body, some from the crust or mantle of another. Thus Figure 7.8 is useful for showing the various environments in which differentiated meteorites formed, but as a model of *the* meteorite parent body, it is as far from the truth as Daly's grand scheme was. Things just were not that simple.

8

The Iron Meteorites
and Pallasites

We owe the largest meteorite in captivity to a man who is far better known as the first to reach the North Pole. In 1894, on the second of several expeditions to explore Greenland, Lieutenant— later Admiral—Robert E. Peary fell under a spell that had beguiled polar explorers since 1818, when Captain John Ross of the Royal Navy found Eskimos using weapons and tools tipped with nickel-iron and heard stories of the "iron mountain" from whence the metal had come. Led to the site on the coast of West Greenland by a hunter, Peary discovered that the "mountain" was really three huge chunks of meteoritic iron: a 59-ton, almost wholly buried mass that the Eskimos called *Ahnighito* or "the Tent," and two smaller masses called "the Woman" and "the Dog."

With the determination that would take him to the North Pole 15 years later, Peary set out to bring the three Cape York irons back to New York City. The job, which his grandson has described in vivid detail (Stafford, 1980), entailed four summers of backbreaking work punctuated by winters of lecturing and lobbying for funds—no small task in those pre-NASA years. But Peary persevered, and in October 1897, *Ahnighito* reached New York (Figure 8.1).

I got to know the first three Cape York irons (several more pieces were found more recently) when they were in the foyer of the Hayden Planetarium at New York City's American Museum of Natural History. Like most other school children who visited the museum, I was awed by the immensity of "the Tent" and fascinated by the ablation pits that dot its dark bronze surface like giant thumbprints. Visitors can see the Cape York irons to still better advantage today in the

113

Figure 8.1 *Ahnighito* ("the Tent"), biggest of the Cape York iron meteorites. (*top*) Members of Peary's party struggle to take the 59-ton meteorite to the coast of Greenland. (Photograph by Lt. Robert E. Peary, negative no. 329141.) (*bottom*) *Ahnighito* arrives at the American Museum of Natural History. (Photograph by Orchard, negative no. 45086; both photographs courtesy of the Department of Library Services, American Museum of Natural History.)

museum's new Arthur Ross Hall of Meteorites. As a long-time fan of *Ahnighito,* I was disappointed to learn that although it is the biggest meteorite ever displayed in a museum, it is not the biggest ever found. That distinction belongs to the Hoba meteorite, which was discovered on a farm in South West Africa in 1920. Hoba measures 9 × 9 × 3¼ feet and weighs about 66 tons. Far too big to move, it still lies where it was found, though pieces of it have found their way to many museums.

Iron meteorites command our attention for several reasons, one of which is their majestic size. Though we know of no other chunks of metal as big as Hoba and *Ahnighito,* one of Western Australia's Mundrabilla meteorites weighs 10 to 12 tons, and the Willamette, Oregon, iron—also in residence at the American Museum of Natural History—weighs 13.5 tons. In contrast, the biggest single piece of a stony meteorite—the Norton County enstatite achondrite, which fell in Kansas in 1948—weighs barely a ton.

Iron meteorites also intrigue us because our forebears used them in their early efforts at metallurgy. The Eskimos' use of the Cape York irons illustrates this, as does the scarcity of iron meteorites in much of the Middle East. As our need for some of the rarer metals that occur in such meteorites outruns terrestrial supplies, we may well turn to the irons again or, more likely, to the asteroids from which they come.

But iron meteorites and the closely related pallasites fascinate us most because they take us to the deep interiors of their parent bodies and bear clues to processes that took place in the Earth billions of years ago and gave our planet its metal core and magnetic field. Irons and pallasites also give us some of the strongest evidence we have for the number and sizes of meteorite parent bodies.

The Widmannstätten Structure

Museum curators and meteorite researchers often get calls from people who think they have found a meteorite. Most of these discoveries prove to be something else—one curator built a large collection of such objects, which he whimsically called "meteor-wrongs"—but exceptions are common enough so that most of us greet such calls with courtesy or even warmth. How can we tell whether a rusty chunk of metal that surfaces in someone's garden is an iron meteorite or a piece of an old plow? A simple test is to shave off a bit of the sample and test it for nickel, which is present in all meteoritic metal. A more elegant test, which works for all but a few iron meteorites, is based on

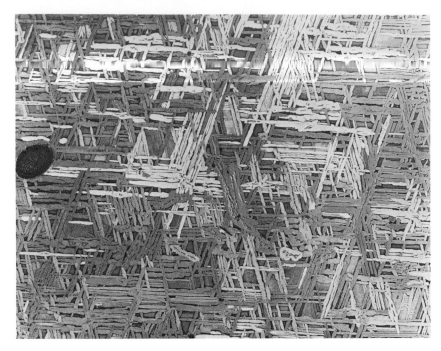

Figure 8.2 Polished, etched slab of the Carbo iron meteorite (medium octahedrite), showing bright kamacite plates separated by darker taenite to form the Widmannstätten structure observed in most irons. The field viewed is 4¾ inches wide. (Smithsonian Institution photograph, reproduced with permission.)

the fact that their metal has a highly characteristic texture: it consists of two nickel-iron alloys arranged in a geometrical pattern (Figure 8.2). This pattern, discovered in 1804 by Count Alois von Widmannstätten in Vienna and William Thomsen in Naples, appears when an iron meteorite is sliced, polished, and etched lightly with a dilute solution of nitric acid in alcohol.

The Widmannstätten pattern both positively identifies most iron meteorites (a few, which contain either a great deal of nickel or very little, lack an obvious pattern) and tells us a lot about their history. To see how this pattern formed and why it is important to our story, we have to take a short detour into another branch of science: metallurgy.

Like the carbon, silica, and olivine discussed earlier in the book, nickel-iron alloys occur in more than one crystalline form. At high temperatures and for high nickel contents, the atoms in such alloys arrange themselves in the pattern shown in Figure 8.3a: they lie at the corners of cubes and in the centers of cube faces. A natural nickel-iron alloy with this "face-centered cubic" structure is called *taenite*. At lower temperatures and for low nickel contents, nickel-iron alloys adopt a second structure, in which each cube of atoms surrounds a central atom (Figure 8.3b). This arrangement is called "body-centered cubic," and the corresponding natural alloy is known as *kamacite*.

By slowly cooling nickel-iron alloys in the laboratory, metallurgists have determined which polymorph occurs at what combinations of temperature and composition. Figure 8.4 summarizes the results of this research. Like the diagram for water-salt mixtures shown in the previous chapter (Figure 7.1), this figure is a map in which composition and temperature substitute for the more familiar coordinates of latitude and longitude. Unlike the salt-water diagram, this one shows only what happens to nickel-iron alloys at temperatures well below melting.

To see how the Widmannstätten pattern developed in iron meteorites, let us follow an alloy of typical meteoritic composition (about one part nickel to nine parts iron) as it cools. Figure 8.4 tells us that at high temperatures, such an alloy (composition a) has the structure and form of taenite. As the temperature falls, the alloy enters the

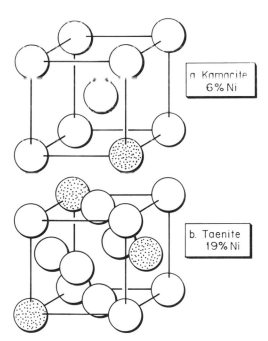

a. Kamacite 6% Ni

b. Taenite 19% Ni

Figure 8.3 Structures of common nickel-iron polymorphs, with atoms spread apart to show their geometrical relationships. Nickel atoms are stippled, iron atoms clear. (a) Body-centered cubic structure of nickel-poor kamacite. (b) Face-centered cubic structure of nickel-rich taenite.

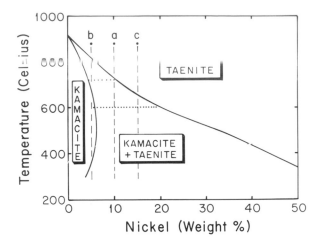

Figure 8.4 Distribution of kamacite and taenite in iron-nickel alloys at various temperatures, based on experimental studies by Romig and Goldstein (1979). Horizontal dotted lines connect the compositions of the two forms of metal at different temperatures. Compositions a, b, and c are appropriate to octahedrites, hexahedrites, and ataxites, respectively.

region marked "taenite + kamacite." When it does so, the taenite structure can no longer hold all of the iron present, and nickel-poor kamacite begins to form within it. At this point, the taenite has the same composition as the bulk alloy (10 percent nickel), the kamacite a more iron-rich composition (slightly more than 4 percent nickel). As the temperature continues to fall, kamacite grows at the expense of taenite, and both minerals become more nickel-rich. For example, at 600° C, taenite contains about 20 percent nickel and kamacite about 6 percent.

Figure 8.4 shows why iron meteorites that contain about 10 percent nickel consist of two nickel-iron alloys, but why do these minerals form the distinctive crisscross pattern that we see in such meteorites? This structure arises because kamacite does not appear at random in the taenite but grows on planes that are oriented like the faces of a regular octahedron (Figure 8.5). This octahedral arrangement of kamacite plates gives us our name for those iron meteorites that have an obvious Widmannstätten pattern: *octahedrites*. Most iron meteorites are octahedrites, because most contain between 6 and 13 percent nickel.

Iron-nickel alloys that contain much less than 10 percent nickel (for example, composition b in Figure 8.4) evolve in much the same way as they cool, but they arrive at a different destination. Because such alloys are iron-rich, the kamacite plates become very thick—so thick that they replace almost all of the taenite. The structure that results consists almost wholly of cubic crystals of kamacite. Since a cube is hexahedral (six-sided), we call nickel-poor iron meteorites that have this structure *hexahedrites*.

It is easy to see why nickel-poor iron meteorites lack the Widmann-stätten structure, but why is it absent from nickel-rich meteorites as well? To understand this, we need to look more closely at what happens when metal cools. For kamacite to form and grow in taenite, nickel and iron atoms have to move through the metal, a process called *diffusion*. Atoms diffuse rapidly at high temperatures; hence kamacite plates grow swiftly at first. However, as the temperature falls, diffusion becomes more and more sluggish, causing the growth of kamacite to slow down and, eventually, stop altogether. Figure 8.4 shows that as their bulk nickel content increases, alloys enter the

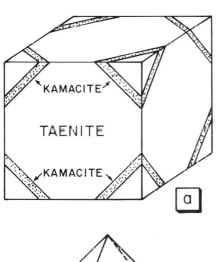

Figure 8.5 Geometrical distribution of kamacite plates in a crystal of taenite (a). These plates parallel the faces of a regular octahedron (b).

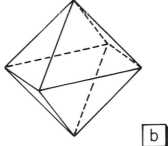

kamacite + taenite region at lower and lower temperatures. Thus alloy b starts to form kamacite at about 800° C, but alloy c does so at 660°. Kamacite appears in the latter alloy at such a low temperature that it has very little time to grow before diffusion stops. Hence the kamacite-taenite intergrowth in a meteorite of composition c is extremely fine-grained—so fine-grained that the meteorite appears structureless to the naked eye. Our name for nickel-rich iron meteorites, *ataxites,* refers to this apparent lack of structure.

Thus all three textural varieties of iron meteorites—the abundant octahedrites and the less common hexahedrites and ataxites—reflect slow cooling of metal with different nickel contents. This conclusion may not seem worth the long discussion that led to it, but it is very important. As we shall see later in this chapter, careful analysis of the Widmannstätten pattern has told us a great deal about the structures and sizes of iron meteorite parent bodies.

Classification of Iron Meteorites

Students of iron meteorites divide them into eight types on the basis of texture: hexahedrites, ataxites, and six kinds of octahedrites with different thicknesses of kamacite plates. This structural classification is of very long standing, and it remains useful for describing newly discovered iron meteorites. But just what do the chemical and structural differences among iron meteorites mean? Did these meteorites form in one parent body or in many bodies? Do they come from centrally located cores or from small pockets of metal that are scattered through the parent bodies like the fruit in raisin bread?

When I examined the question of the number of chondrite parent bodies, I based my answer on three pieces of evidence: first, small but sharp compositional differences among the eight or nine chondrite groups; second, the prevalence of one-group breccias; and third and most compelling, differences of oxygen isotopic composition from one group to the next. These three arguments will be used again in Chapter 9 when I discusss the various kinds of achondrites, and I shall add a fourth criterion, age, that is particularly useful for working out the genealogy of those igneous meteorites.

Most of the criteria that we use to work out associations of silicate-bearing meteorites are of little or no use for the irons. Though many iron meteorites bear the scars of shock events, few are breccias whose components might tell us which kinds of irons formed close together and which formed far apart. Fewer contain minerals that we can use to detect oxygen isotopic differences or to establish different times of

formation. Even traditional chemical analyses tell us little about relationships among iron meteorites, for they report just a handful of the
more abundant elements: iron, nickel, cobalt, sulfur, and small
amounts of carbon and phosphorus.

To subdivide and interpret the iron meteorites, we rely principally
on *trace elements,* which can be defined for our purposes as those
chemical constituents whose abundances are so low—commonly a
few atoms per million or billion iron atoms—that they do not appear
in traditional chemical analyses. To study these elements, chemists
irradiate samples with neutrons to convert the trace elements to
radioisotopes; they then use the decay of these isotopes to tell which
elements are present and in what abundances.

To understand how trace elements help us interpret the iron
meteorites, we need to know something about the way in which a
nickel-iron liquid crystallizes. When molten nickel-iron cools, the
metal that forms at each stage of crystallization contains less nickel
than the associated liquid. Hence fractional crystallization produces a
mass of solid metal that is iron-rich at its base and progressively
enriched in nickel upward (Figure 8.6). John Lovering showed in

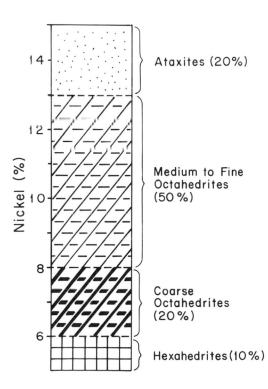

Ataxites (20%)

Medium to Fine
Octahedrites
(50%)

Coarse
Octahedrites
(20%)

Hexahedrites (10%)

Figure 8.6 Cross-section of a
differentiated nickel-iron core
formed by fractional crystallization of a liquid composed
of 11 percent nickel and 89
percent iron by weight. The
distribution of various kinds
of iron meteorites in this core
is based on calculations by
John Lovering (1957).

1957 that fractional crystallization of a liquid with the composition of an average iron meteorite (about 11 percent nickel) can produce solids that span the observed range of iron meteorite compositions. This simple process can even account for the observation that octahedrites are far more abundant than nickel-poor hexahedrites and nickel-rich ataxites.

How would trace elements be distributed in a differentiated metal core? As we saw in the discussion of layered igneous intrusions, the trace elements in a magma distribute themselves among crystallizing minerals in accordance with Pauling's rules. Those ions that fit comfortably in the structure of olivine, oxidized nickel for example, enter that mineral and follow it to the bottom of the intrusion. Other elements that fit in feldspar, for example strontium, follow that mineral. Still others, including some that are commercially important, find no homes in the major igneous minerals and accumulate in the remaining liquid, from which they may eventually crystallize to form ore deposits. As a result of this process of sorting trace elements according to their affinity for various minerals, samples taken from various levels in a layered intrusion describe distinctive trends of abundance for different trace elements. Thus nickel is abundant near the bottom of such an intrusion and decreases more or less regularly upward; strontium shows the opposite tendency. By studying trends for these and other trace elements, geologists can work out the crystallization history of an intrusion in considerable detail.

The rules that govern trace element distribution in a crystallizing magma work just as well in a crystallizing nickel-iron liquid. In fact, trace element trends in a metallic core are simpler than those observed in igneous rocks, since there are fewer kinds of crystals to compete for trace elements. Thus we would expect elements that prefer solid nickel-iron to liquid to be carried down by settling crystals and concentrated near the center of the core. Elements with the opposite preference should remain in the liquid, to be concentrated upward in the core.

Figure 8.7 shows some simple relationships between iridium and nickel that might result from fractional crystallization of iron-nickel liquids with various starting compositions and histories. The first point to note in this diagram is that different scales are used for nickel (percent) and iridium (parts per million); the second is that iridium varies far more widely than nickel. Much of the value of trace elements lies in the fact that fractional crystallization affects them much more strongly than it affects the major elements.

The data points in Figure 8.7 are not scattered across it but are gathered together in elongate fields or trends. Trend A is what one would expect to find in random samples of the differentiated core shown in Figure 8.6. Because iridium prefers solid metal to liquid, it is most abundant in the nickel-poor cumulates that lie deep in the core; its abundance decreases upward as the nickel content increases. Fields B, C, and D in Figure 8.7 show different trends of iridium abundance that might result from fractional crystallization of liquids with different initial compositions—for example, richer (B, C) or poorer (D) in nickel. They also show that iridium-nickel diagrams and other trace-element diagrams can tell us a lot about crystallization history. Trend A shows the wide chemical variation that results from very efficient separation of crystals and liquid. In contrast, the modest variation evident in B indicates very slight fractional crystallization.

For our present purposes, the important question is whether the iron meteorites lie on one trend in trace-element diagrams or on many trends, that is, whether they formed from one liquid or from many liquids. Early studies of trace elements in iron meteorites were inconclusive on this point. When Lovering and his co-workers analyzed 88

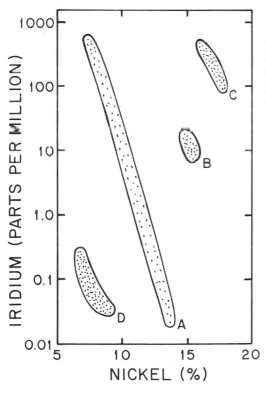

Figure 8.7 Hypothetical distributions of iridium and nickel in the core shown in Figure 8.6 (A) and in cores that are, on average, richer (B, C) or poorer (D) in nickel.

irons for gallium and germanium in 1957, they found that most of them fell into four groups on a diagram that compared these two elements, but that the four groups appeared to lie on one line. Lovering and his colleagues regarded the groups as important enough to label (I to IV), but they were more impressed by their alignment, which suggested that they all came from one differentiated core.

The simple picture that emerged from this early trace-element study vanished when other researchers examined more elements in more iron meteorites. Between 1967 and 1978, John Wasson and his colleagues at UCLA analyzed more than 500 irons for gallium, germanium, and iridium. Figure 8.8 summarizes their data for iridium and nickel. This figure and analogous diagrams for germanium and gallium make it clear that the iron meteorites do not fall on one trend but constitute at least 12 chemical groups. Some of these groups overlap in Figure 8.8, but they are distinct in diagrams for other elements. In naming these iron groups, Wasson and his co-workers used the Roman numerals that Lovering had devised, adding letters to subdivide the four original groups. In cases where two groups appeared to be related, they showed this by using two letters. Thus groups IIIA and IIIB, which are similar in many respects and appear to intergrade, appear as the field IIIAB in Figure 8.8.

Because there is no obvious way for crystal-liquid differentiation to shift meteorites from one chemical group to another, Wasson and his colleagues concluded that the 12 to 16 groups defined by trace elements sample at least a dozen parent bodies. As startling as this conclusion is, it tells only part of the story: the UCLA researchers also found that about 14 percent of the iron meteorites they analyzed fell in *none* of the known iron groups but were scattered wildly on trace-element diagrams. These unclassified irons may represent perhaps 50 more groups, samples of which seldom reach the Earth.

The groups that Wasson defined on the basis of trace-element data differ in other respects as well. For example, Figure 8.8 shows that some groups consist almost wholly of nickel-poor (IIA and IIB) or nickel-rich (IVB) meteorites, that is, hexahedrites or ataxites. Others, for example IIIA and IIIB, include only octahedrites. There are also mineralogical differences: most iron meteorites contain more or less iron sulfide (troilite) and iron-nickel phosphide (schreibersite), but members of some groups contain other minerals as well. For example, two groups of irons contain olivine, pyroxenes, and other silicates, which have yielded most of the meager data we have on the ages of iron meteorites. Other groups contain minerals—for example the chromium nitride carlsbergite—that are known from nowhere else.

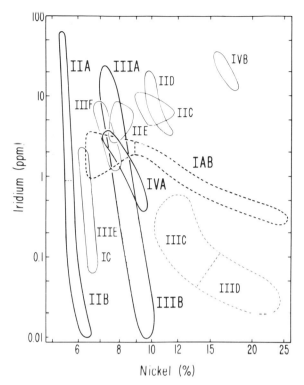

Figure 8.8 Distribution of nickel and iridium in various chemical groups of iron meteorites. (After Scott and Wasson, 1975, reproduced by permission of the authors and the American Geophysical Union.)

Thus detailed chemical study of the iron meteorites has produced a picture that is even more complex than the one we saw for chondrites. And as we shall see, cooling rates complicate that picture still further.

Cores or Raisins?

A question that trace-element distributions do not answer is whether the iron meteorites formed in centrally located cores or in small pods of metal that were scattered through their parent bodies. The answer to this question lies in the rates at which iron meteorites cooled. Irons that formed at about the same depth in a parent body—in its core, for example—should have cooled at very similar rates. Irons that formed at various depths, and hence were insulated by different amounts of overlying rock, should have cooled at very different rates.

But how can we tell how fast an iron meteorite cooled? The answer lies in the phenomenon that explains the very fine grained Widmannstätten structure found in nickel-rich ataxities: sluggish diffusion of iron and nickel at low temperatures. If diffusion were rapid at all temperatures, an octahedrite would consist of homogeneous, nickel-rich taenite and homogeneous, nickel-poor kamacite; we would expect a series of microprobe analyses across the two minerals to show a nickel distribution like that in Figure 8.9b. In fact, the microprobe sees a very different pattern of nickel variation as it makes its way across an octahedrite (Figure 8.9c): the edge of each taenite region is nickel-rich as expected, but the nickel content drops inward, passes through a minimum value near the center of the taenite, and then rises toward the opposite edge. The reason for this M-shaped nickel profile is that during the last stages of kamacite growth, diffusion was too sluggish to carry nickel into the centers of taenite regions.

In the early 1960s, shortly after the microprobe became available for use by meteoriticists, two young scientists, Joseph Goldstein and John Wood, independently suggested that the shape of the M profile in taenite reflects the rate at which the metal cooled and can be used to estimate that rate. A very flat M, similar to the ideal nickel distribution shown in Figure 8.9b, testifies to extremely slow cooling—slow enough for diffusion to produce nearly homogeneous taenite. The sharp, deep M profile in the left-hand taenite plate shown in Figure 8.9c reflects very rapid cooling, while the shallower M in the right-hand plate of part c reflects an intermediate cooling rate. By calculating the shapes of nickel profiles that should result for various

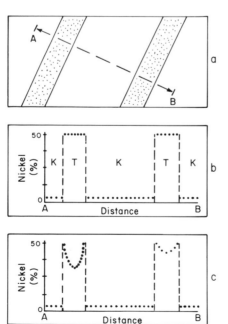

Figure 8.9 Growth of the Widmannstätten structure at various rates of cooling. (*a*) Kamacite plates (clear) and taenite (stippled) in a hypothetical octahedrite, showing the path traversed by the microprobe in parts b and c. (*b*) Distribution of nickel anticipated for infinitely slow cooling or perfectly efficient diffusion. (*c*) Nickel distributions expected for a high (left-hand plate) or a moderate cooling rate (right-hand plate).

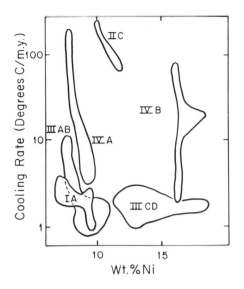

Figure 8.10 Cooling rates and nickel contents of various iron meteorite groups. (From Dodd, 1981.)

bulk meteorite compositions and cooling rates, Goldstein and Wood were able to translate the microprobe's observations into cooling rates for various kinds of octahedrites. Their method, subsequently simplified and modified by themselves and by others, has been applied to a great many iron meteorites and to other metal-bearing meteorites as well.

Figure 8.10, which compares the nickel contents and cooling rates of several groups of iron meteorites, shows very clearly that the irons had varied thermal histories. Some groups, notably IIIA, IIICD, and IIC, show little variation of cooling rate, suggesting that each group formed in one mass of metal—perhaps, though not necessarily, a central core. Other groups, in particular IVA and IVB, have widely varied thermal histories, which suggest that they formed in smaller masses—raisins—scattered throughout their parent bodies.

Twenty years ago meteoriticists argued almost as vehemently about whether iron meteorites formed in cores or raisins as they did about the number and sizes of parent bodies. This issue has now faded into history. As is often true in science, the answer to this "either/or" question has proved to be "both."

Pallasites

Although there are only 35 pallasites, all but two of them finds, these olivine stony-irons are important as a bridge between the iron and stony meteorites. They are also among the handsomest of rocks, with beautifully formed, golden crystals of olivine set in gleaming metal like topaz in a fine ring. Evidently earlier residents of the Americas also appreciated the pallasites, for one of them—Hopewell Mounds—was discovered at a prehistoric burial site, and pieces of others were found so far apart that we believe primitive man transported them over long distances.

The Thiel Mountains pallasite, an Antarctic meteorite shown earlier (Figure 3.3), is typical of the class in that it consists of about two-thirds magnesian olivine, one-third metal, a bit of troilite and schreibersite (iron-nickel phosphide), and little else. Readers with unusually sharp eyes may see two kinds of metal in Figure 3.3, for Thiel Mountains, like other pallasites, contains kamacite and taenite arranged in a Widmannstätten pattern. This pattern is much more obvious in Figure 8.11, which shows a metal-rich portion of Brenham, an unusually inhomogeneous pallasite.

Long before meteoriticists had access to the microprobe and neutron-activation analysis, the textural similarity between the metal in pallasites and that in iron meteorites led them to suspect that the two kinds of meteorites are very closely related. These modern tools have confirmed that suspicion: most of the pallasites have trace-element distributions and cooling rates that closely resemble those observed

Figure 8.11 The Brenham, Kansas, pallasite, showing an unusually metal-rich region with a well-developed Widmannstätten structure. (Smithsonian Institution photograph, reproduced with permission.)

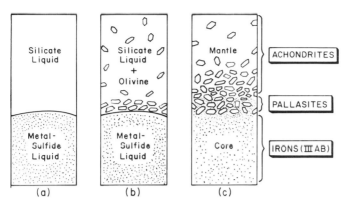

Figure 8.12 Development of pallasites and related iron meteorites from a chondritic liquid. (*a*) Metal-sulfide and silicate liquids separate, with the former sinking because of its high specific gravity. (*b*) Fractional crystallization of the silicate liquid produces a cumulate of settled olivine crystals just above the metal-silicate boundary. (*c*) As the metal-sulfide liquid solidifies, the nickel-rich residual liquid is drawn upward into the olivine cumulate to produce pallasites.

for one of the largest iron groups, IIIAB (see Figures 8.8 and 8.10). The few exceptions suggest that pallasites, like iron meteorites, sample more than one parent body.

There is still some debate about just how pallasites formed, but we have little doubt that they sample one or more boundary regions between masses of metal and silicates—environments analogous to the Earth's core-mantle boundary. One plausible process for the formation of pallasites is shown in Figure 8.12. Molten chondritic material separated into two liquids: a metal-sulfide core and a silicate mantle (Figure 8.12a). As the mantle cooled, olivine crystallized early and settled out to form a cumulate just above the core-mantle boundary (Figure 8.12b). How liquid metal from the core moved into the olivine cumulate is less clear (Figure 8.12c). One possibility is that silicate liquid between the olivine crystals moved upward as basaltic magma, and molten metal simply replaced that liquid, just as soda replaces air when you suck on a straw. Another is that the mantle shrank as it cooled, squeezing molten metal out of the core and into the olivine-rich layer.

We conclude, then, that most of the pallasites are samples of the core-mantle boundary of the body that housed group IIIAB iron meteorites. The few pallasites that have unusual compositions may have formed in similar environments, but in different parent bodies.

Clearly, the IIIAB parent body has sent us generous samples of its core and lower mantle. Has it sent us bits of its upper mantle and crust as well? We shall return to that question when we examine the igneous achondrites in Chapter 9.

History: Theme and Variations

Twenty years ago it seemed possible to describe the history of iron meteorites quite simply: (1) a chondritic parent body started to melt; (2) the resulting iron-nickel-sulfur liquid trickled inward to form a molten core; (3) that core solidified, with iron-rich crystals sinking to the center through a more nickel-rich liquid; (4) at a late stage in crystallization, nickel-rich liquid from the core invaded the parent body's lower mantle to produce those rocks that we call pallasites.

At least one group of iron meteorites—that labeled IIIAB in Figures 8.8 and 8.10—played this theme more or less as written, but many other groups played quite elaborate variations on it. The fact that some groups have varied cooling rates suggests that they formed as dispersed pockets of metal rather than central cores. One group, the IAB irons, has a peculiar iridium distribution (see Figure 8.8) and other properties that lead us to think these meteorites were never entirely molten. Evidently these peculiar, silicate-bearing irons sounded just the first notes of the iron meteorite theme before they froze in a form that they have maintained for 4.5 billion years.

A glance at Figures 8.8 and 8.10 shows us that the various iron groups differ in average composition as well as history: Some consist entirely of iron-rich and some of nickel-rich meteorites, hexahedrites and ataxites respectively. Clearly, these groups formed from liquids with different compositions. How can we explain these differences? As we saw in Chapter 7, differences in the availability of oxygen could produce nickel-rich and nickel-poor liquids from one type of chondritic parent material. If that material melted under oxygen-rich (oxidizing) conditions, it would produce a nickel-rich liquid and a solid residue of iron-rich silicates (Figure 7.9b). If it melted under oxygen-poor (reducing) conditions, the liquid would be much more abundant and richer in iron, and the solid residue would consist of iron-poor silicates (Figure 7.9c).

No doubt some of the chemical differences among iron meteorite groups are due to different degrees of oxidation or reduction. Indeed, the reason for our suspicion that some nickel-poor irons, specifically those of group IIAB (Figure 8.8), are related to the enstatite chondrites and achondrites is that all three kinds of meteorites record strongly reducing conditions. However, some chemical differences among the iron meteorites—for example, the high iridium contents of some and the low iridium contents of others—cannot be explained in this way. It is clear that iron meteorites formed from several different kinds of chondritic material.

Before moving on to consider other differentiated meteorites, I shall conclude this chapter on the irons and pallasites with a brief look at what we know—and still do not know—about two important aspects of these meteorites: when they formed and what kinds of parent bodies they came from.

Few iron meteorites contain minerals that can be dated by conventional means. In fact, we have radiometric ages for only two groups, and we are not sure just what those ages mean. The most reliable ages now in hand come from a small group of irons (IIE) that contain feldspar and can thus be dated by the very precise rubidium-strontium method. Some of these meteorites have ages that are similar to those of chondrites—about 4.55 billion years. Others have younger ages, down to a minimum of 3.8 billion years for the Kodaikanal meteorite. Taken at face value, these dates suggest that the parent body of IIE meteorites started to melt very shortly after the sun formed and was still molten, at least in part, 700 million years later.

We are no longer surprised by evidence that the meteorite parent bodies became very hot very early, but it is hard to understand how any of them could have stayed hot for as long as 700 million years. This is particularly puzzling in the case of irons, whose parent bodies were, it seems, very tiny indeed. It is unfortunate that IIE is a very small group, for its peculiar history invites further study.

The other group of silicate-bearing irons (IAB) lacks feldspar, but iodine-xenon ages of other silicates and troilite in these meteorites are very high—close to those obtained for chondrites. It appears that these meteorites, like some of the IIE irons, formed very early in the history of the solar system.

We know nothing about the formation ages of the other iron meteorites, most of which lack silicates altogether. Fortunately, this problem has attracted the attention of nuclear chemists, who are exploring isotopic systems based on elements—such as palladium and silver—that occur in meteoritic metal. If this work is successful, it will strengthen what is now a very weak link in our knowledge of iron meteorites.

The final question that must be asked about the varied objects that produced iron meteorites and pallasites is how big they were. We saw in Chapter 6 that we can use both pressure-sensitive minerals and cooling rates to determine how deeply a meteorite was buried in its parent body. If that meteorite formed at the center of the parent body, its depth of burial equals the body's radius; if not, it gives us only a minimum radius for its parent body. Laboratory experiments tell us that the Widmannstätten pattern cannot form at pressures greater than about 12,000 atmospheres. Therefore, no octahedrite comes from the center of a body with a radius greater than about 500 miles. This fact alone rules out parent bodies as big as the moon. Unfortunately, it is impossible to use pressure arguments to narrow further the range of permissible body sizes, because few irons contain pressure-indicative minerals.

However, most irons do contain kamacite and taenite, and we can use the nickel distributions in these minerals to calculate cooling rates. These cooling rates, in turn, can be used to estimate depths of formation. When I made such calculations a few years ago (Dodd, 1981), I concluded that those iron groups that sample cores formed in bodies with radii between 120 and 170 miles—the size of moderately large asteroids. Calculations for other groups yielded similar sizes, which are, of course, minima.

As noted in Chapter 6, recent laboratory experiments on iron-nickel alloys have changed this picture dramatically. When we take into account the minor elements in iron meteorites, the calculated cooling rates increase sharply and the sizes we infer for parent bodies drop just as sharply. It now appears that almost all iron meteorites cooled at rates of at least 100° C per million years. These revised cooling rates are startling and a bit unsettling, for they imply parent bodies less than 12 miles in diameter for those iron meteorites that sample central cores. Larger bodies are possible for iron groups that sample "raisins," but even these meteorites appear to have formed in rather small asteroids. Thus parent bodies that we once pictured as at least as big as the moon have shrunk, in just 20 years, to objects about as big across as Manhattan Island is long.

The iron meteorite parent bodies are unlikely to "shrink" further, for the new cooling-rate calculations have eliminated the principal causes of uncertainty in the old ones. In fact, I suspect that the tiny parent bodies now in view will prove to be too small: it is very hard to see how such gravity-driven processes as fractional crystallization and core formation could take place in tiny bodies with insignificant gravitational fields. Perhaps our assumption that similar cooling rates for a group of irons indicate that if formed in a central core is incorrect, since all this really tells us is that a group of meteorites formed at about the same depth.

Once thought to be the simplest of meteorites, the irons have proved on careful study to be surprisingly complex. Just how many parent bodies they represent is a fascinating question. Taken at face value, the many irons that do not fit in our present trace-element classifications—about 14 percent of those studied to date—suggest that the Earth has sampled bits of perhaps 50 more bodies during the last few hundred years. Do these unclassified irons come from objects that are just beginning to favor us with samples, or from parent bodies that made most of their contributions long ago?

The ancient iron meteorites from Antarctica may answer this question. It will be interesting to see whether some of the orphan irons will find their families among meteorites that fell thousands, even millions, of years ago.

9

Achondrites and Their Parent Bodies

Two major objectives of meteorite research are to identify those differentiated meteorites that formed in the same body and to use them to reconstruct and identify that body. The meteorite associations that we have identified to date, some of them well-established, others speculative, are listed in Table 9.1.

Using a handful of meteorites to reconstruct their parent body is a hard job, rather like trying to reassemble Humpty Dumpty after the king's men have run off with most of the pieces and his horses have trampled the rest. How do we go about it? Most of the arguments used in earlier chapters to show that meteorites come from *different* bodies can be turned upside down to identify groups of meteorites that formed in the *same* body. Consider breccias, for example: we saw in Chapter 4 that the prevalence of one-group chondritic breccias indicates that the various kinds of chondrites formed far enough apart to preclude much mixing among them; this in turn suggests that these groups of meteorites occupied at least eight parent bodies. But as we shall see in the next section, the fact that some other kinds of meteorites consistently occur together in breccias leads us to conclude that they formed very close together, probably in the same body.

Chemical arguments also work both ways. Sharp chemical differences among the three classes and eight (or nine) groups of chondrites suggest that they formed in different bodies. On the other hand, evidence that enstatite chondrites lie on the same chemical trends as enstatite achondrites suggests that these two groups formed in the same body and are related by a process of partial melting (Chapter 6).

Table 9.1 Associations of igneous meteorites

Association	Member	Environment	Body
Eucrite	Eucrites	Volcanic/intrusive	Asteroid
	Diogenites	Intrusive (cumulate)	(Vesta)
	Howardites	Surface (regolith)	
	Mesosiderites	Surface (regolith)	
SNC	Shergottites	Volcanic/intrusive	Planet
	Nakhlites	Intrusive (cumulate)	(Mars)
	Chassignite	Intrusive (cumulate)	
Pallasite-	Pallasites	Core-mantle boundary	Asteroid
iron	IIIAB irons	Core	
Enstatite	En achondrites	Mantle(?)	Asteroid
meteorites	En chondrites	Crust	
	IIAB irons	Core	

Note: All associations but the last are now widely accepted.

As we saw in Chapter 8, similar chemical relationships suggest that the IIIA and IIIB irons and most of the pallasites also shared a body.

We can also use oxygen isotopes to both separate meteorites and bring them together. As observed in Chapter 5, the different oxygen isotopic distributions found in various kinds of silicate-bearing meteorites are the strongest evidence we have that they formed in different parent bodies. Similar oxygen relationships suggest, on the other hand, that two or more groups of meteorites are related.

Ironically, the use of oxygen isotopes, one of the best tools that we have for establishing relationships among meteorites, also provides a good example of the hazards involved in using just one kind of evidence for this purpose. Lunar and terrestrial rocks differ in many respects and obviously formed in different bodies, but they lie on one mass-fractionation line in an oxygen diagram. Like the other criteria that we have used to link meteorite groups—similar shock histories or cooling rates, similar ages or degrees of oxidation—isotopic similarity of two groups of meteorites only suggests that they formed together; it does not prove that they did.

To build a case for a meteorite association, we must line up as many kinds of evidence as we can, in the hope that all of them will point in the same direction. In two of the cases that we shall consider in this chapter, this is so. The third example shows that, despite heroic efforts by many scientists, some kinds of meteorites are still orphans, with no obvious connections to other branches of the meteorite family tree.

Case 1: The Eucrite Association

One would never guess by looking at a typical eucrite and a diogenite that they are related, for even though both meteorites are breccias made up of bits of igneous rock, the igneous materials that comprise them are very different. Eucrites consist of basalt-like rocks composed chiefly of calcium-poor pyroxene (pigeonite) and plagioclase feldspar, with small amounts of metallic iron, troilite, and one or more silica minerals. These rocks have varied histories—some are volcanic (see Figure 7.3, top), others intrusive rocks like those of the Palisades (Figure 9.1, bottom)—but all of them crystallized from basalt-like magma on or near the surface of their parent body. In contrast with the eucrites, diogenites (Figure 9.1, top) consist almost entirely of crystals and fragments of one mineral—iron-magnesium pyroxene (specifically, bronzite)—with very small amounts of plagioclase, chromite, troilite, and metal. The few diogenites that escaped brecciation are much coarser-grained than eucrites and look very much like the pyroxene cumulates that we find in the Stillwater Complex and other layered terrestrial intrusions.

Clearly eucrites and diogenites are very different rocks. What makes us think they are closely related? The most obvious evidence is that fragments of eucrites and diogenites are the principal components of two kinds of breccias: the howardites and the metal-rich mesosiderites (see Figure 7.10). G. T. Prior noted this relationship some 70 years ago and concluded from it that howardites are mixtures of bits of eucrites and diogenites, and that mesosiderites are mixtures of the same silicate materials plus metallic nickel-iron.

We have seen so many old ideas swept away by new data that it is refreshing to encounter an exception. Although detailed studies of howardites and mesosiderites have shown that they contain small amounts of other materials as well, several pieces of evidence—chemical gradations among eucrites, diogenites, howardites, and mesosiderites (Figure 9.2); their similar oxygen isotopic compositions (Figure 9.3); and their similar ages, very close to those of chondrites—tell us that Prior was right. These four groups of differentiated meteorites are closely related and probably formed in the same parent body. I therefore call them, collectively, the *eucrite association*.

A lunar connection? In the decade before Apollo 11 touched down on the moon's Sea of Tranquillity, many meteoriticists doubted that meteorites could make the long trip from the asteroid belt to the Earth on the tight schedule indicated by their cosmic ray exposure ages. They therefore looked for more plausible sources of meteorites:

Figure 9.1 Thin sections of intrusive members of the eucrite association, viewed in transmitted light. (*top*) The Johnstown diogenite, a breccia composed almost entirely of iron-magnesium pyroxene crystals and fragments, which show varied colors (here shades of gray) in doubly polarized light. Field of view is about 1 inch wide. (Photograph by Thane Bierwert, courtesy of the Department of Library Services, American Museum of Natural History; negative no. 126959.) (*bottom*) The Serra de Magé eucrite, composed of coarse-grained pyroxene (light gray) and plagioclase (striped). Though chemically similar to Ibitira (Figure 7.3, top), this rock crystallized more slowly below the surface of the parent body and has a cumulate texture. Doubly polarized light reveals the distinctive texture of plagioclase, a result of twinning, and tiny plates of calcium pyroxene (white) in low-calcium pyroxene (dark crystal above twinned plagioclase). Field of view is about 1/16 inch wide. (From Dodd, 1981.)

comets, for example, and the moon. The moon appeared to be a particularly attractive source for the basalt-like eucrites and shergottites, for most scientists interpreted the huge, dark lunar maria as impact basins filled with basalts. Thus when Apollo 11 left for the moon in 1969, most of us guessed it would return with basalts, and many looked forward to a bonanza of eucrites and, perhaps, shergottites.

We were half right. Mare Tranquillitatis and the other maria that the Apollo and Luna landers sampled do contain basalts, and those rocks do resemble the eucrites and shergottites in chemical composition and mineralogy. However, there are also many differences be-

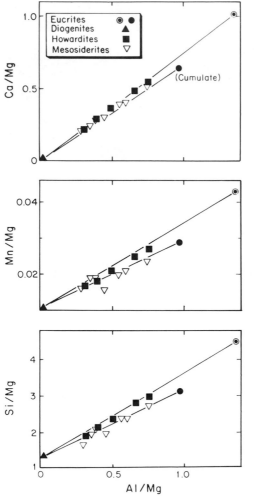

Figure 9.2 Chemical variations among members of the eucrite association, based on data from Louis Ahrens and his associates at the University of Capetown. Mean compositions of cumulate and noncumulate eucrites are distinguished. In this and other chemical comparisons, howardites and the silicate portion of mesosiderites appear to be mixtures of eucritic and diogenitic materials, with minor additional components.

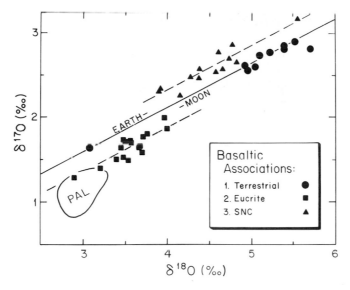

Figure 9.3 Three-isotope oxygen diagram for members of three basaltic associations and for pallasites (PAL). The Earth-moon mass-fractionation line is shown for comparison. The data, compared with standard mean ocean water (SMOW), are from R. N. Clayton and T. Mayeda (1978 and 1983).

tween lunar and meteoritic basalts, the most significant of which are their oxygen isotopic compositions and ages. As we have seen, all lunar rocks hug the Earth's mass-fractionation line on an oxygen diagram. In contrast, most of the eucrites and their associates lie well below that line (Figure 9.3). Eucrites and lunar mare basalts also formed at quite different times, the former about 4.5 billion years ago and the latter between 3 and 4 billion years ago. These differences and subtler chemical distinctions make us certain that the eucrites and their associates do not come from the moon. The surprisingly low ages of shergottites—about 1.3 billion years—lead us to the same conclusion for those meteorites as well.

Though lunar rocks and those of the eucrite association are not related, they formed in much the same way and show us that basaltic volcanism, which I have called a leitmotif of planetary evolution, has been such since the very beginning of the solar system's history. The shergottites make the same point.

There is another parallel between lunar rocks and those of the eucrite association. When the Apollo astronauts set out to sample the moon, they were disappointed to find that a shovel was more useful for the job than a rock pick: solid bedrock as we know it does not exist on the moon, whose airless surface is mantled by a thick blanket of *regolith,* a fine-grained, soil-like breccia produced by aeons of pounding of the lunar rocks by large and small meteoroids. Evidently the eucrite parent body also has a thick regolith, for careful microscopic comparison of howardites and samples from the lunar surface has shown them to be very similar, down to the presence in both of micrometeorite craters and noble gases implanted by the solar wind. Since the eucrite fragments in howardites bear textural and mineralogical evidence that they crystallized at widely different rates and at various depths, we believe that the crust of the eucrite parent body was shattered and stirred—or "gardened," to borrow a colorful term from the lunar scientists—to a depth of a mile or more. The varied potassium-argon ages of shocked eucrites suggest that this gardening went on for a very long time, probably 1.5 billion years. Thus both lunar rocks and their meteoritic equivalents record long histories of impact reworking on their respective parent bodies.

The eucrite parent body. If the eucrites and their associates do not sample the moon, where do they come from? The high abundance of solar wind gases in brecciated members of the association indicates that they spent a very long time in the inner solar system and narrows our choice of sources to the terrestrial planets and asteroids. High crystallization ages—close to 4.5 billion years—make the latter objects more likely, since volcanism started late in bodies as big as the moon and continued over long periods of time. The eucrite parent body, like the tiny bodies that housed chondrites, became hot very early and cooled very fast.

Melting experiments on eucrites and diogenites and detailed studies of their trace elements also point to a small eucrite parent body. They indicate that the magmas that produced these igneous rocks formed by partial melting of chondrite-like mixtures of olivine, pyroxene, and feldspar rather than the high-pressure equivalents of these minerals. Thus melting took place at low pressures—not impossible in a big body, but far more likely in a small one.

A final point in favor of an asteroidal parent body for eucrites and their associates is that an excellent candidate is at hand. The reflectance spectrum of asteroid 4 Vesta resembles spectra measured for eucrites and their associates very closely (see Figure 5.6); Vesta's specific gravity is also appropriate. Since we know of no other large asteroid that meets these specifications, it is likely that Vesta, one of the biggest asteroids, is the ancestral home of the eucrite association.

A pallasite-eucrite association? We are so used to dividing meteorites that it is a heady experience to lump together four groups in one parent body. Can we go further and say that the pallasites and the IIIAB irons formed in Vesta too? The oxygen isotopic data for pallasites and the eucrite association make this grand association seem plausible, because the two kinds of material appear to share a mass-fractionation line (Figure 9.3). However, a problem emerges when we try to imagine what a eucrite-pallasite parent body would look like.

Consider the body shown in Figure 7.8. Clearly, it has niches for all of the groups of meteorites under consideration: the eucrites and their associates sample its brecciated upper crust, pallasites its core-mantle boundary, and irons its core. The presence of intrusions in the crust explains our observation that some of the coarser eucrites (Figure 9.1, bottom), the diogenites (Figure 9.1, top), and some other bits of rock that we find in howardites and mesosiderites are crystal cumulates. The problem with all this—a severe one—is that we know of no meteorites that sample this body's olivine-rich mantle. There is no obvious way around this difficulty. We can adjust the thickness of the mantle by, for example, making different assumptions about the fate of oxygen in the parent body (see Figure 7.9), but all reasonable assumptions lead to a mantle that makes up between 40 and 60 percent of the body by volume. If the eucrite association and the pallasites come from the same body, we have managed to obtain many samples of its crust and core without sampling its mantle—a feat that deserves to be called magic!

This argument alone persuades us that the eucrite association and the pallasites sample two different bodies. Vesta's nearly spherical shape supports this conclusion, for it is hard to see how impacts could mine the asteroid's core and core-mantle boundary without leaving the rest of the body in jagged pieces. If the eucrites and their relatives come from Vesta, then the pallasites and IIIAB irons probably sample one of the many asteroids whose reflectance spectra tell us that they consist of mixtures of metal and silicates.

Case 2: The SNC Meteorites

When nearly all of the numbers in a set of data are much the same, scientists tend to distrust exceptions. In 1969, meteoriticists were so used to seeing meteorite ages between 4.5 and 4.6 billion years that they were skeptical when the Nakhla achondrite (Figure 7.7b) yielded an age of 1.4 billion years. Because that age was based on an isotopic system—potassium-argon—that is notoriously easy to reset by heating, most researchers assumed that Nakhla's very young age signifies late metamorphism of a much older igneous rock.

We were wrong. Though no one suspected it at the time, Nakhla's crystallization age, confirmed later by many other dating methods, was the first hint that at least one meteorite parent body was still alive and producing magma long after the moon ceased to do so. Recent studies of the other nakhlites, the shergottites (Figure 7.3b), and the unique olivine achondrite Chassigny (Figure 7.7, top) have shown that all of these igneous rocks crystallized during a short period of time about 1.3 billion years ago. These studies, well summarized by McSween (1984) have also shown beyond reasonable doubt that all of them come from the same body, certainly a planet and probably Mars.

The SNC association. The shergottites, nakhlites, and chassignites, which add up to about a dozen meteorites, seem at first glance to have very little in common. The basalt-like shergottites resemble eucrites in mineralogy and texture and, like them, are samples of shallow layered intrusions and volcanic flows. In contrast, the nakhlites and Chassigny are cumulate rocks, the former composed mainly of the calcium-rich pyroxene augite and the latter almost wholly of olivine. These meteorites also have very different shock histories: the shergottites experienced impacts that were vigorous enough to convert their plagioclase to glass and cause local melting, whereas Chassigny records mild shock and the nakhlites record virtually none.

The features that unite these meteorites, in addition to their common age, are subtle but significant. For example, all contain relatively iron-rich silicates and iron oxides, which testify that they formed in a rather oxygen-rich environment (see Figure 7.9). In this respect and in their content of small amounts of water-bearing minerals, they differ sharply from eucrites and their associates and resemble terrestrial igneous rocks. The shergottites, nakhlites, and Chassigny also share an oxygen mass-fractionation line that differs slightly from both the eucrite and Earth-moon lines (Figure 9.3). These features and their common crystallization age make us all but certain that these three kinds of meteorites formed in the same body; thus we call them the shergottite-nakhlite-chassignite association or, more concisely, SNC achondrites (pronounced "snick").

Size of the SNC parent body. Like the eucrites and diogenites, the SNC meteorites are products of partial melting of chondritic parent material and fractional crystallization of the resulting basaltic magmas. In both cases the products are volcanic and shallow intrusive igneous rocks, the latter including cumulates.

Historical parallels between the two associations end there. If the SNC parent body possesses a regolith like that on the eucrite parent body and the moon—material similar to howardites and mesosiderites—we have not sampled it. Moreover, the timing of igneous activity was quite different in the two bodies. The eucrites and diogenites formed about 4.5 billion years ago, during an episode of igneous activity that lasted no more than 100 million years—just a blink of an eye in geological terms. Although impacts stirred the regolith on the eucrite parent body for 1.5 billion years, samples of that regolith appear to include no younger magmatic rocks. We are forced to conclude that the eucrite parent body's active life began and ended very early in the history of the solar system. The SNC meteorites also record just one brief episode of igneous activity, but it took place more than 3 billion years later. Moreover, detailed isotopic and trace-element studies of the SNC meteorites tell us that their parent body underwent two earlier episodes of differentiation: one 4.5 billion years ago, the other between 2.8 and 1.4 billion years ago. In contrast to the short-lived eucrite parent body, the SNC body had a long, complex internal history.

These historical differences between the two parent bodies indicate that they employed different sources of heat. The early and brief episode of melting recorded in eucrites and diogenites, like the short metamorphic history of ordinary and enstatite chondrites, requires a heat source that was present at the beginning of the solar system and vanished very shortly thereafter. As noted in Chapter 6, short-lived radioactivity—specifically aluminum-26—is the most likely of several candidates.

The very late episode of melting that produced the SNC meteorites clearly had nothing to do with such temporary heat sources as short-lived radioactivity or an active sun. Rather, it must reflect either internal heating by long-lived radioisotopes or surface heating by meteoroidal impact. The first alternative commits us to a large parent body, for even a big asteroid would lose heat to space fast enough to balance that produced by slow radioactive decay. In fact, evidence that the SNC parent body was still producing magma almost 2 billion years after lunar volcanism ceased suggests that that body is bigger than the moon and is in fact a planet.

The alternative interpretation of the SNC meteorites—that they formed on an asteroid, in puddles of shock-generated magma—arose because of what once seemed to be a fatal shortcoming of the big-body model: the problem of ejecting material from a planet without shocking it beyond recognition. Shergottites are intensely shocked, but the nakhlites and Chassigny are not. The discovery of bits of modestly shocked lunar rock in Antarctica softens this objection to a big body but does not quite remove it. The Antarctic finds indicate that recognizable meteoritic material can make its way from the moon to the Earth, but they do not prove that virtually unshocked samples could make a longer trip from a bigger body. The problem of delivering SNC meteorites remains a serious objection to a planetary source for such meteorites.

The impact-melting model has problems of its own, however. One is that SNC meteorites show little resemblance to the shock melts that we find beneath terrestrial and lunar impact craters. In particular, these meteorites lack the abundant fragments of shattered wall rock that are typical of shock melts. Crystal cumulates like the nakhlites and Chassigny are also rare in shock melts, most of which crystallized much too swiftly to allow large crystals to form and settle. A more serious objection to the impact-melting model is that it accounts for only the episode of melting and differentiation that produced the SNC achondrites, not the earlier episodes that affected their parent body. That body's long, complex internal history is very hard to explain with any heat source other than long-lived radioactivity.

These shortcomings of the asteroid model are largely responsible for a recent shift of scientific opinion toward a planetary source for the SNC achondrites, but there is also some positive evidence for such a source. For example, some studies of trace elements in shergottites suggest that the region that produced their parent magma contained at least one high-pressure mineral, garnet, and detailed examination of the cumulate shergottites suggests that they reflect crystal settling in a substantial gravity field. Both of these observations point to a large parent body.

The case for Mars. An important reason for the shift of opinion toward a big parent body for SNC achondrites is that several bits of evidence point specifically to one planet, namely Mars. Mars has two advantages over other planets that might have yielded the SNC meteorites: it is close to the Earth, and its small size and thin atmosphere minimize the problem of ejecting samples from its surface. (Venus, whose surface rocks are known to include basalts, is less promising than Mars because it is much bigger and has a very thick atmosphere.) Moreover, studies of the distribution of impact craters on the Martian surface suggest that some of the planet's many volcanoes were active at about the time when the SNC meteorites formed.

Some properties of the meteorites themselves also point to Mars. For example, shergottites chemically resemble some samples of the Martian soil that were analyzed by the Viking landers, and all of the SNC meteorites share two properties—oxidation and enrichment in water and other volatile constituents—with that soil. Finally, one of the most intensely shocked shergottites contains noble gases—argon, krypton, and xenon—whose isotopic compositions resemble those found in the Martian atmosphere. The only plausible explanation for this observation is that the meteorite trapped these atmospheric gases during shock melting.

In summary, a great deal of evidence, of many types and from many sources, points to a planet, specifically Mars, as the source of the shergottites, nakhlites, and Chassigny. Just how these meteorites escaped from Mars remains unclear, but most meteoriticists are now quite sure that they did.

Case 3: Ureilites

We have accounted for most of the silicate-bearing differentiated meteorites listed in Table 7.1. Some of the remaining groups are represented by only one (angrite, siderophyre) or two meteorites (lodranites). Of these rare meteorites, only Angra dos Reis has been studied in detail. We know that this pyroxene cumulate formed at about the same time as the eucrites and suspect that it formed in a large body, but we have no idea how it is related to other meteorites. We know even less about the other rare meteorites.

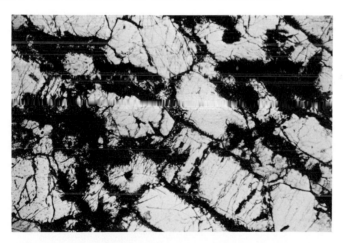

Figure 9.4 Thin section of the Kenna ureilite, showing elongate crystals of olivine and calcium-poor pyroxene (pigeonite) in roughly parallel orientation and separated by black, carbon-rich material that contains diamond and lonsdaleite. The photograph, about 3/32 inch wide, was taken in transmitted light. (From Dodd, 1981.)

The last group of differentiated meteorites, the ureilites, is too big and intriguing to overlook; it includes more than a dozen of the most bizarre rocks ever collected on or off the Earth. The ureilites appear at first glance to be just one more kind of crystal cumulate, in this case composed of olivine and the calcium-poor pyroxene pigeonite (Figure 9.4). What distinguishes them from all other cumulates is the black material between the silicates, which contains abundant carbon and includes both of the high-pressure forms of that element: diamond and lonsdaleite.

The silicate minerals in ureilites bear abundant evidence of intense shock, and we are sure that their diamond and lonsdaleite record that process rather than high pressure in a big parent body. We are sure of very little else about these odd rocks. Though it now appears that their carbon was present in the parent magma, we have no idea how, when, or where that magma formed. Oxygen isotopic data and trace elements suggest that ureilites are related to the carbonaceous chondrites, but the nature of the relationship is to a large extent unknown. Despite a great deal of work by many researchers, the ureilites remain the most puzzling of meteorites—orphans indeed.

This chapter completes the survey of the kinds of meteoritic material that have struck the Earth in the recent past. The remainder of the book will focus on two important and fascinating questions: First, what do meteorites tell us about the birth and infancy of the solar system? Second, what role did meteorites play in the history of the Earth and its inhabitants?

10

Meteorites and
the Early Solar System

All human beings, from Nubian tribesmen to Nobel laureates, share an inborn fascination with beginnings. This fascination, expressed first in the child's question, "Where did I come from?" leads geologists to hunt for the oldest rocks, anthropologists for the earliest man, and physicists for the most primitive nuclear particle. Cosmologies and creation myths differ in sophistication, but they answer the same human urge: to know how our story began.

Modern attempts to explain how the solar system began take three directions. Planetary scientists, the tribe of which meteoriticists are a small clan, describe the modern solar system, looking for clues to how things were from how they are today. Astronomers examine the radiation that streams from space, trying to identify and describe those stars that are today as the sun was 4½ billion years ago. Finally, astrophysicists integrate the findings of these observational scientists, using physical theory to span the immense gaps in our knowledge and create reasonable hypotheses for the birth and infancy of the sun and its family.

These three approaches work hand in hand, and none is of much use without the others. However, studies of meteorites play a crucial role, for they are the only rocks we have that formed in the early solar nebula and survived to tell us of their experiences. Indeed, each major advance in our understanding of meteorites helps astrophysicists draw a tighter rein on speculation and inch closer to an accurate picture of the early solar system.

Framing the Problem

Any successful interpretation of the solar system must explain its properties, of which the following are the most important.

1. It consists of a rather small, middle-aged star, nine planets, dozens of satellites, thousands of asteroids, and millions of comets. Most of these objects, comets excluded, revolve around the sun in the same direction and in almost the same plane.

2. Most of the solar system's mass (99.9 percent) resides in the sun, but most of its motion—which we express as angular momentum—resides in the planets.

3. The planets are of two types, which we call inner (or terrestrial) and outer (or giant). The inner planets—Mercury, Venus, Earth, Mars—are relatively small, dense objects composed of rock and metal. The outer planets, with the exception of Pluto, are big, low-density objects composed of gases and ices. The asteroids, which thinly populate the gap between Mars and Jupiter, are much smaller than the terrestrial planets, but like them consist of rock and metal in various proportions.

4. Though the compositional differences between the inner and outer planets are most striking, there are also chemical differences among the terrestrial planets that are reflected in their mean specific gravities. When we correct for the effect of pressure, which compresses minerals and may convert them to denser polymorphs, the specific gravities of these planets decrease more or less regularly with distance from the sun (Figure 10.1). The only conspicuous exception to this pattern is the moon, which lies at the same solar distance as the Earth but is much less dense. Both this fact and chemical analyses of lunar rocks indicate that the moon contains less iron and a larger proportion of nonvolatile elements than the Earth.

5. The inner planets differ in mass as well as specific gravity. As Figure 10.1 shows, their masses increase with distance from the sun from Mercury to the Earth, but the fourth planet—Mars—is small. Even if we assume, generously, that the asteroid belt once contained as much mass as Mars contains now, it is obvious that the region between the Earth and Jupiter is mass-poor.

Although there are many other properties of the solar system that we might consider, they tell us little about its history. For example, if we assume that the asteroid belt once contained enough material to constitute a planet, then there is a regular mathematical relationship among planetary orbits that is called the Titius-Bode law. This relationship has intrigued astronomers for centuries, but they now attach

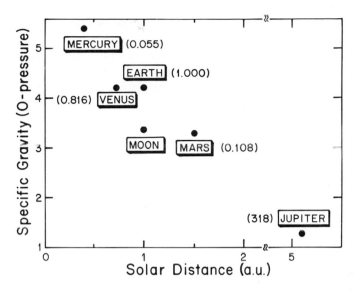

Figure 10.1 Zero-pressure specific gravities of the inner planets and Jupiter in relation to their solar distances. Numbers in parentheses are masses of other planets relative to the Earth.

little significance to it: it may just represent the closest possible spacing of stable planetary orbits.

With these observations in mind, we can move on to consider the evidence that we use to explain them, starting with that contributed by astronomy.

Evidence from Stars

Astronomers are short-lived creatures who ponder long-lived objects. Because they can witness only a tiny fragment of each star's history, they have to build a picture of stellar evolution from glimpses of many stars, in much the same way that an editor builds a movie from many film clips. Fortunately, the astronomers have abundant material to work with—there are almost 50 billion stars in our galaxy alone—and stars appear to follow fairly simple patterns from birth to death. Let us examine those patterns briefly.*

Stars form from *interstellar clouds,* unusually dense patches of gas and dust that are typically about 10,000 times as massive as our sun.

---

*The reader who wants a fuller treatment of the subject will find one in John Wood's excellent book *The Solar System* (1979).

For somewhat obscure reasons that appear to be related to the spiral structure of galaxies like our Milky Way, such clouds contract and break up into many smaller, slowly rotating *cloud fragments,* each of which is a potential star. The cloud or a part of it may become a star cluster. After a fragment separates from its parent cloud, it continues to shrink under the influence of its own gravitation, drawing in more gas and dust as it does so. Gravitational energy provided by the accreting material heats the center of the fragment, a process that becomes more effective as added dust insulates the center against losses of heat to space. As the cloud fragment's internal temperature rises, so does its gas pressure. Eventually, the pressure becomes great enough to balance the inward pull of gravitation, and the fragment ceases to collapse. At this point, perhaps 10,000 years after its separation from the cloud, the fragment becomes a *protostar,* which continues to gather dust and gas from the surrounding cloud, alternately heating and cooling as it passes through several episodes of contraction and expansion. During this period, the protostar shows wide and rapid changes in brightness.

Protostars can achieve very high internal temperatures as they contract (the physical principle involved—heating a gas by compressing it—is the same one employed in the diesel engine), but until they become hot enough for nuclear reactions to occur, their only source of energy is gravitation. When their central temperatures exceed about 10 million degrees, such reactions come into play: hydrogen nuclei unite to form helium in a process that astronomers call hydrogen burning. The onset of this thermonuclear reaction, which takes place only in protostars that are at least a tenth as massive as our sun, marks the transition from protostar to star. Our sun crossed this important frontier about 10 million years after its parent cloud fragment formed.

To this point, cloud fragments and protostars have been treated as if they were static, spherical objects. In fact, rotation plays an important role in determining their shapes and the number of stars that they will form. Each cloud fragment starts its career with a quantity of motion, its *angular momentum,* which is proportional to its speed of rotation and the square of its radius. Because angular momentum is conserved (remains constant) as the fragment evolves, changes in its radius are offset by changes in its speed. Thus a cloud fragment accelerates as it contracts to become a protostar, for the same reason that a whirling ballerina spins faster when she draws her extended arms toward her body.

Cloud fragments acquire different amounts of angular momentum from their parent clouds. A fragment that possesses little angular momentum will never rotate very fast, and most of its mass will therefore be concentrated near its center (Figure 10.2A). A fragment with more initial momentum will spin faster to produce a flattened, disk-like central mass (Figure 10.2B), two or more mass concentrations (Figure 10.2C), or even a doughnut-shaped ring. Thus cloud fragments that possess little angular momentum are likely to evolve into single protostars and, eventually, single stars, while fragments with much momentum may produce two or more protostars and a binary or multiple star system.

In contrast to protostars, whose energy output (luminosity) and surface temperature vary widely as they expand and contract, most stars produce thermonuclear heat about as fast as they lose it to space. Hence each has an almost constant luminosity and temperature related to its mass. This relationship among luminosity, surface temperature, and mass is shown in Figure 10.3, which is based on observations of a great many stars. All but the oldest and youngest stars in this figure lie on a diagonal band that astronomers call the *main sequence,* with the more massive stars at the bright, hot ("blue") end of the band and the less massive ones (the sun included) at the dim, cool ("red") end.

Stars spend most of their lives on the main sequence, entering it shortly after they are born and leaving it only when they start to die. It took the sun only 10 million years to evolve from an interstellar cloud fragment to a star. In contrast, it has been on the main sequence for 4½ billion years. Fortunately for life on Earth, whose survival depends on a stable source of heat, the sun will remain there for several billion more; but when the hydrogen in its core is exhausted,

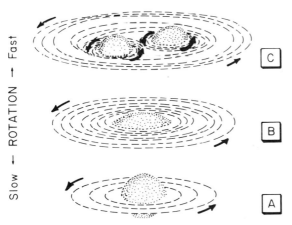

Figure 10.2 Effect of various rates of rotation on the forms of cloud fragments and protostars. (After John A. Wood, *The Solar System,* copyright 1979, p. 159. Adapted by permission of Prentice-Hall, Englewood Cliffs, N.J.)

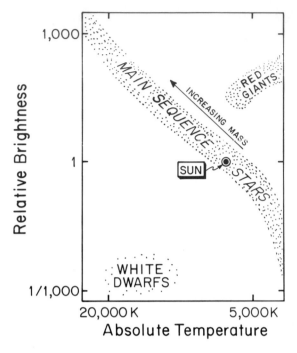

Figure 10.3 Stellar evolution diagram, showing relationships among mass, relative brightness (sun = 1), and absolute temperature (Kelvin). (After John A. Wood, *The Solar System*, copyright 1979, p. 135. Adapted by permission of Prentice-Hall, Englewood Cliffs, N.J.)

our star will move off the main sequence into the region of red giants (Figure 10.3), then will shed its outer portions and become a white, and eventually a black, dwarf. If humanity manages to avoid subjecting itself to an earlier catastrophe, it will come to judgment at the red giant stage, when the bloated sun will first roast the Earth and then swallow it.

If the sun were much larger than it is, it would come to a quicker and more spectacular end, because massive stars burn hydrogen very rapidly and thus have briefer careers on the main sequence. When their cores become depleted in hydrogen, such stars pass through an elaborate series of thermonuclear reactions and explode to form supernovas and, finally, black holes. It is likely that a nearby supernova produced the short-lived radioisotopes recorded in meteorites and possible that it also triggered the cloud collapse that led, ultimately, to the solar system.

Young suns. For our present purposes, the ultimate fate of the sun is less interesting than its early history. What does astronomy tell us about protostars and young stars, objects that are today where the sun was 4½ billion years ago? Such objects are hard to study because of their obscuring mantles of dust and gas, but astronomers have examined enough of them to note two peculiar properties: first, they show dramatic variations of brightness on time scales as short as days; second, they expel large amounts of their substance in the form of streams of ionized gas, a phenomenon called T-Tauri mass ejection. (Both of these features of protostars—enhanced luminosity and a very strong solar wind—were mentioned in Chapter 6 as alternatives to short-lived radioactivity for heating meteorite parent bodies.) Evidence that protostars experience mass loss also explains why most astrophysicists believe the cloud fragment that produced the solar nebula was about twice as massive as the modern sun.

Studies of young stars also tell us that most of those that resemble the sun in other respects are members of binary or multiple systems, implying that most interstellar cloud fragments possess considerable angular momentum. Our solar system, which includes just one star, appears to be an exception. The accepted explanation for this singularity is that the sun formed from a cloud fragment with an unusually modest amount of angular momentum.

Recent studies suggest that multiple star systems are even more common than astronomers had thought. Some stars that have no visible companions show a curious tendency to wobble, which suggests that they are influenced by the gravitational pull of one or more cool, and hence invisible, objects nearby. In 1984 careful study of infrared radiation from such a wobbling star, van Biesbroeck 8 in the constellation Ophiucus, revealed an odd, faintly glowing companion—a body much bigger than the planet Jupiter but small enough so that it is not burning hydrogen. This "brown dwarf" has touched off a semantic controversy—is it a gigantic planet or a failed star?—and generated a great deal of excitement in the astronomical community. If van Biesbroeck 8's dim companion is a failed star and if such brown dwarfs are common, then a multiple system may well be the *typical* result of star formation, making our solitary sun all the more unusual. Perhaps it is time to consider the possibility that the giant planets are remnants of a second, failed protostar.

Do any other young stars have properties which suggest that they have or are forming planetary systems? The best candidate identified to date is Vega, one of the brightest stars in the northern sky. Data returned by the Infrared Astronomy Satellite (IRAS) in 1983 suggest that a swarm of solid objects is in orbit around Vega. These objects are much larger than grains of interstellar dust and have the same mass ratio with their star—about 1:1000—that the planets in our solar system have with the sun. It appears that Vega either possesses a planetary system or is evolving one; astronomers cannot now choose between these alternatives.

The recent observations of van Biesbroeck 8 and Vega are major steps forward in astronomers' attempts to identify and characterize other planetary systems. As satellites like IRAS and the Space Telescope carry our instruments above the Earth's obscuring atmosphere, we will no doubt find other suns in the making. But astronomy can take us just so far. The new data from Vega show both that science's potential and its limitations: they tell us that the solid material around Vega is coarser than interstellar dust, but not whether that material includes objects that are big enough to be called planets. For the foreseeable future, astronomy will give us only a very fuzzy picture of planetary systems—something like the view of a baseball game that you would get if you watched it on a three-inch television screen from across a large room. To see more, we must move closer to the game. Chondrites permit us to do so.

Evidence from Chondrites

Chondrites have a great deal to tell us about events in the swirl of gas and dust that we call the primitive solar nebula, for evidence that they contained short-lived radionuclides tells us that they formed early in its history, while the sun was still a protostar.

Chondrites give us three important clues to the history of solid material in the nebula. First, they tell us that the formation of the solar system involved one or more processes that coarsened interstellar dust. Studies of that dust suggest that most of its particles are no more than a thousandth as big as the period at the end of this sentence. In contrast, most meteoritic chondrules are about as big as that period, and many are much bigger. The coarsening processes, which we can call, loosely, "chondrule formation," played an important role in the evolution of nebular material. Chondrites also tell us that the ingredients of planets were isotopically inhomogeneous. We have already seen that the isotopic composition of oxygen varies significantly among different kinds of meteorites (Figure 6.2) and, apparently, between two planets, the Earth and mars (Figure 9.3). The components of chondrites show even wider isotopic variations; they extend to elements other than oxygen; and they are, at least in part, of presolar origin. Finally, chondrites tell us that the ingredients of planets were chemically inhomogeneous. Though all chondrites chemically resemble the sun, the various classes and groups are compositionally distinct (see Figure 4.6). Obviously, some process or processes of chemical differentiation affected nebular material before it accumulated to form the meteorite parent bodies and, ultimately, the planets.

Each of these three clues to nebular history is important enough to discuss in some detail.

Chondrule formation. If the science of meteoritics has a central mystery, it is the origin of chondrules: the tiny, spherical or rounded bits of igneous material that are present in most chondrites and the predominant material in many of them. Although de Bournon first described such objects almost two centuries ago and many researchers have studied them since then, we are still unsure just how, where, and from what kind of material chondrules formed.

One reason for the mystery surrounding chondrules is that they are remarkably varied objects. Though all of them formed from molten or partly molten material and most consist principally of olivine, pyroxene, or both, they have varied forms and textures that imply varied histories. Some chondrules are simply frozen droplets of silicate liquid (Figure 10.4, top left and bottom left), but others are irregular bits of igneous rock (Figure 10.4, top right), and many bear evidence of more than one episode of melting, metamorphism, or both (Figure 10.4, bottom right). Indeed, chondrules are so diverse that some workers prefer to restrict the term "chondrule" to the frozen droplets; they call the other objects rock fragments or clasts. I shall call all of them chondrules here, since I believe that the two kinds of objects are very closely related.

Chondrules also puzzle us because we have never seen them form. A few chondrule-like objects are scattered through the lunar regolith and in material sprayed from impact craters on Earth, but we know of no process that produces chondrules in the proportions observed in chondrites. Moreover, many of the processes that have been invoked to explain them are impossible to simulate in the laboratory. Thus we have to interpret chondrules without the aid of analogues, either natural or artificial.

Of the many explanations of chondrules that have been proposed, only one—explosive volcanism—strikes all modern researchers as implausible. The others are summarized in Table 10.1, which has been arranged according to the kind of starting material or precursor that each model envisions: hot solar gas, dust, or preexisting rock.

The idea that chondrules formed from droplets of liquid that condensed from solar gas goes back to Henry Sorby, who described them, vividly, as "fiery rain." Recent proponents of condensation have recognized that at the very low gas pressures that prevailed in the primitive solar nebula, mineral grains should have condensed rather than liquid. Thus all modern condensation models include plausible means—enhanced pressure, excess dust, or supercooling, for example—to get around that problem. However, a far bigger problem plagues the various condensation hypotheses. Through meticulous microscopic study and microprobe analysis, Hiroko Nagahara and Ermanno Rambaldi have shown that many chondrules, perhaps most of them, contain bits of unmelted material—relics of what were clearly solid precursors. This crucial observation has persuaded most researchers that chondrules formed by melting of preexisting solids rather than by condensation of nebular gas.

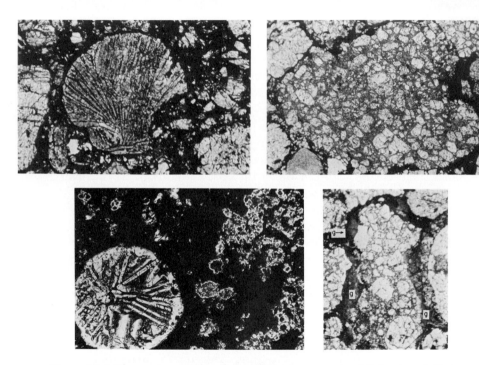

Figure 10.4 Photomicrographs of chondrules in transmitted light. (*top left*) A drop-formed chondrule in the Kota-Kota enstatite chondrite, composed almost wholly of pyroxene plates in a fan-like arrangement. The chondrule is 0.5 mm in diameter. (*bottom left*) A drop-formed chondrule in the Allende carbonaceous chondrite, composed of sets of parallel plates of olivine (light) and reddish-brown glass (dark). The chondrule, 0.5 mm in diameter, is accompanied by an unmelted aggregate of olivine crystals. (*top right*) An irregular, clast-formed chondrule in the Manych LL-group ordinary chondrite. The chondrule, which is 3.3 mm (1/8 inch) long, consists chiefly of well-formed olivine crystals (light) and brownish glass (dark), with scattered grains of metal and troilite (black). Its texture, similar to that of porphyry but on a much finer scale, is called microporphyritic. (*bottom right*) A remelted microporphyritic chondrule in the Manych chondrite. The chondrule, 2.5 mm long, resembles the one shown above in texture and mineralogy, but it is surrounded by brown glass (g) that signifies remelting. The photograph was overexposed to show the glass. (All but the bottom right photograph are from Dodd, 1981.)

COLLEGE OF THE SEQUOIAS LIBRARY

We have the following title which was
purchased at your suggestion. It will
be in the workroom for the next day or
two--then added to our collection. If
you would like our assistance in locating
it, please bring this notice to the
library workroom and either of us will be
pleased to help you.

Marla Decker - Susie Myers

Date _____

Table 10.1 Theories of chondrule formation

Precursor	Process	Location	Time
Nebular gas	Condensation	Nebula	Early
Interstellar dust	Melting by lightning, impact, or friction	Nebula	Early
Igneous rock	Impact melting	Parent body	Late
Differentiated igneous rock	Impact melting	"Grandparent bodies"	Late

What those solids were like, and how and where they melted, are far less certain. One possibility is that the immediate precursors of chondrules were fluffy aggregates of fine-grained interstellar dust—something physically similar to the dust balls that form under beds. Those workers who prefer such precursors invoke nebular processes to melt them. Among the melting processes that are in favor today are high-velocity impacts between dust balls, lightning, and frictional heating of interstellar dust at the edge of the dusty nebula. According to such interpretations, chondrules formed before nebular material began to accrete to build the meteorite parent bodies and, ultimately, the planets. Thus they are very primitive objects indeed.

Some kind of dust model would now win a solid majority among meteoriticists, but a few of us find such models hard to reconcile with two aspects of chondrules. One is that they vary quite widely in composition, from olivine-rich (Figure 10.4, bottom left) to pyroxene-rich (Figure 10.4, top left). In view of the immense number of minute dust grains that must come together to form a typical chondrule, one might expect all chondrules that formed from dusty precursors to have the same composition—that of the dust.* I believe that the immediate precursors of chondrules were much coarser-grained than interstellar dust, and the observations of Rambaldi and Nagahara bear me out: some of the unmelted grains that they found in chondrules are thousands of times as coarse as typical interstellar dust. The other shortcoming of dust models is that many chondrules have gone through one or more cycles of metamorphism, deformation, or both, suggesting that they were buried in a body, metamorphosed, and then exhumed to start the cycle again. Such histories

---

* If this argument is not quite clear, picture a bag that contains 500 black marbles and 500 white ones. If you blindly withdrew marbles 100 at a time, most of your samples would contain about 50 black and 50 white marbles.

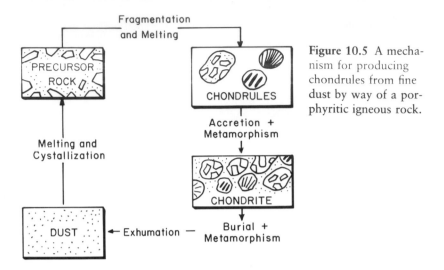

Figure 10.5 A mechanism for producing chondrules from fine dust by way of a porphyritic igneous rock.

seem to me to be absolutely inconsistent with the formation of chondrules in the open nebula. They point, rather, toward a relationship between chondrule formation and parent bodies.

These inadequacies of dust models have led me to propose the interpretation shown in Figure 10.5. I picture chondrules as the products of repeated impacts before and during the growth of the chondrite parent bodies, and I suggest that some impacts were vigorous enough to produce puddles of liquid as much as several inches across. The igneous rocks that crystallized from these puddles—much coarser-grained than the original dust—were shattered by later impacts and, in some cases, remelted. The chondrule shown in Figure 10.4, top right, a type that is very common in ordinary chondrites, may be a rounded fragment of such a "puddle rock"; the one shown in Figure 10.4, bottom right may be such a fragment that was remelted during a later impact.

Although my interpretation of chondrules is well out of the mainstream today, it is less radical than some. For example, Robert Hutchison of the British Museum has observed that some chondrules have chemical compositions which suggest that they were derived from strongly differentiated igneous rocks. This observation has led Hutchison to revive an interpretation that Harold Urey proposed many years ago: that chondrules are bits of differentiated objects—what Urey called "grandparent bodies." Although I disagree with this theory, I share Hutchison's opinion that at least some chondrules formed late rather than early in the progression from dispersed dust to parent bodies and planets.

In summary, chondrules are witnesses to one or more processes in the primitive solar nebula, but they speak a language that we do not yet fully understand. Until we are certain just where, when, and how they formed, these enigmatic objects will continue to tell us less than they know about the history of the solar system. But the news about chondrules is by no means all bad. Although meteoriticists still differ about specific processes of chondrule formation, our conclusion that the starting material was solid is a stride forward. Condensation models imply that the inner solar system was once very hot—hot enough to vaporize and homogenize interstellar dust and erase all evidence of its previous history. On the other hand, evidence of solid precursors suggests a cooler nebula, one that may have preserved some of the dust from which it formed. The next two sections show that such ancient relics did indeed survive, though apparently not in chondrules.

CAI's. We saw earlier that most chondrites consist of chondrules, metal and troilite grains, and fine-grained matrix material in various proportions. Although this description fits ordinary and enstatite chondrites quite well, it falls short for the carbonaceous chondrites. Many such meteorites, in particular those of the CO and CV groups, contain another kind of high-temperature material that consists chiefly of minerals—calcium, aluminum, and titanium oxides and silicates—that are rare in or absent from chondrules. These objects, some of them droplets but most of them unmelted aggregates of crystals, are called calc-aluminous inclusions or CAI's (Figure 10.6). CAI's were first identified in 1968 and sprang to prominence a year later when research teams at the Cambridge and Washington branches of the Smithsonian Institution discovered that they are common in the Allende meteorite. They are also abundant in many other carbonaceous chondrites but are rare in or absent from ordinary and enstatite chondrites.

For three reasons, CAI's have played a very important role in recent discussions of our solar system's history. First, the minerals of which they are composed are extremely refractory: they form only at very high temperatures and thus would be among the first solids to condense from a cooling solar gas. This fact and the observation that some CAI's are unmelted led the Cambridge researchers to suggest that they are aggregates composed of primitive nebular condensates. Second, CAI's show unusually wide variations of isotopic composition for oxygen and many other elements, and much of that variation cannot be explained in terms of familiar processes in the solar nebula. Finally, the chemical compositions of CAI's suggest that they are one

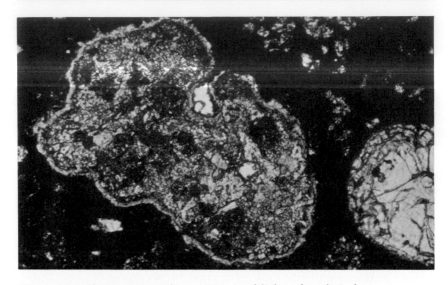

Figure 10.6 Photomicrograph, in transmitted light, of a relatively coarse-grained calc-aluminous inclusion (CAI) in the Felix carbonaceous chondrite (Group CO). The aggregate, 0.7 mm long, consists of several calcium-aluminum silicates and oxides and appears to be unmelted. (From Dodd, 1981.)

of the materials that moved about in the nebula to give the various groups of chondrites their different compositions.

The basis for the suggestion that CAI's are primitive nebular condensates can be seen in Figure 10.7, which summarizes the sequence in which minerals would condense from a thin, cooling solar gas. Compounds of calcium, titanium, and aluminum would condense first, followed in order by metallic iron, pure magnesium silicates (the olivine forsterite and the pyroxene enstatite), and calcium feldspar. If these early crystals remained in the cooling gas, they would react with it at lower temperatures, the calcium feldspar becoming a calcium-sodium feldspar, the olivine and pyroxene acquiring oxidized iron. At still lower temperatures, sulfur would react with some metallic iron to produce troilite, and water would convert olivine and pyroxene to hydrous silicates. Finally, at very low temperatures, water, ammonia, and methane would condense as ices. Although pressure affects the order of appearance of minerals slightly (metallic iron and forsterite can, for example, change places in the sequence), Figure 10.7 presents a reasonably accurate picture of the sequence of condensation in the solar nebula.

In describing this picture, it has been assumed that minerals that condensed early remained in the gas after they formed, an assumption analogous to letting a liquid crystallize completely in a closed pot (Chapter 7). The solids that result must, of course, have the same composition as the original gas. It should be obvious from what was said earlier about fractional crystallization that if we remove all or part of the solids as they form, we can produce condensates with a wide range of compositions. As we shall see, one way to account for compositional variations among meteorites and planets is to assume that they formed from solids that separated from nebular gas at various temperatures.

We can see from Figure 10.7 that one way to produce CAI's would be to cool solar gas to about 2200° F and extract the solids, for example by letting them clump together. Unfortunately, this is not the only way to make them, for we can also read Figure 10.7 in reverse: if we heated chondritic material to 2200° at low pressure, we would vaporize everything else and leave a solid residue composed of the same refractory compounds that we find in CAI's. At the moment, meteoriticists are sharply divided between these alternatives: one camp pictures condensation from a pervasive, hot gas; the other pictures high-energy events—impact, lightning—vaporizing dust to leave residues of refractory minerals. I prefer the second interpretation because some CAI's contain mineral grains with varied isotopic compositions—a result that would be most unlikely if these minerals condensed from a common gas cloud. Such CAI's, like the unmelted relics in chondrules, suggest that the solar nebula was never hot enough throughout to melt or vaporize all of the dust that it inherited from its parent interstellar cloud.

Isotopic anomalies. We have seen that many processes—mass fractionation, radioactive decay, interactions with solar wind or cosmic

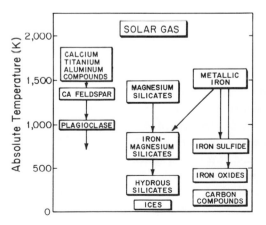

Figure 10.7 Sequence of condensation of minerals from a gas of solar composition at a pressure of about 1/10,000 atmosphere. Arrows show reactions between early condensates and gas to produce minerals that are stable at lower temperatures.

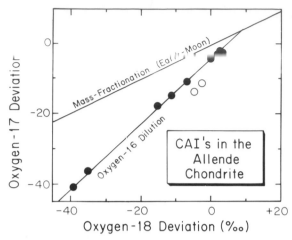

Figure 10.8 Isotopic composition of oxygen in CAI's in the Allende carbonaceous chondrite. Chondrules in the same meteorite show a similar but shorter trend of O-16 enrichment. The two open circles mark CAI's that lie off the trend and show isotopic anomalies for many other elements as well as oxygen.

rays—can produce unusual isotopic distributions in meteorites. If we cannot explain the isotopic composition of a sample with such familiar processes, we call it anomalous and must look for other causes for it.

Since isotopes are synthesized in stars and, in particular, in supernovas, we are certain that the isotopic composition of interstellar matter varies widely in space and time. Thus material from elsewhere in the Milky Way or from beyond it would carry an isotopic signature that would allow us to distinguish it from normal solar system material. This fact led meteoriticists to hunt for isotopic anomalies with great enthusiasm in the 1960s and early 1970s. Unfortunately, the results of this early work were frustrating: most of them were negative, the remainder ambiguous. Skepticism about the existence of isotopic anomalies prevailed until 1973, when Robert Clayton and his co-workers at the University of Chicago discovered that oxygen isotopes vary in meteorites in ways that cannot be explained by mass fractionation—the only plausible source of variation for these stable isotopes. Clayton noted that chondrules, CAI's, and other materials in Allende and other carbonaceous chondrites fall on a 45-degree line in an oxygen diagram (Figure 10.8), suggesting that they contain mixtures of two kinds of oxygen: a "normal" solar-system component and a second component composed of almost pure O-16. He pointed out that the O-16 component could not have formed through any process in the solar nebula, but it might result from destruction of the heavier oxygen isotopes in a supernova.

Because any process that affected oxygen would probably affect other light elements as well, Clayton's group and others looked for other isotopic anomalies in those materials—CAI's—that show the most dramatic enrichments in O-16. Most CAI's proved to lack anomalies in magnesium, silicon, and calcium, but a few—including the two that lie below the carbonaceous chondrite line in Figure 10.8—showed anomalies in all of these elements and many others as well.

Detailed interpretation of the isotopic anomalies in meteorites has proved to be extremely difficult, in part because the processes that produce elements in stars are very complex and poorly understood, but also because new anomalies have appeared faster than astrophysicists and nuclear chemists could explain the old ones. However, it is now clear that the solar nebula was contaminated, very early in its history, by one or more injections of material from beyond it. A supernova is the most plausible source for such material and for the very short-lived aluminum-26, plutonium-244, and iodine-129 that were present in the early solar system. Indeed, it is possible that shock waves generated by that supernova provided the mysterious "push" that caused an interstellar cloud to collapse, starting the chain of events that led to the sun's birth. Thus we may owe the solar system's existence, and our own, to the violent death of an old star.

At the very least, the survival of isotopic anomalies in meteorites confirms the testimony of chondrules that the primitive solar nebula was cool enough to preserve solid mineral grains. Chondrites, long accepted as our best windows on the early solar system, give us an extraordinarily clear view: one that reaches far beyond the sun and its family to the nebula from which they came.

Chemical variation. The last aspect of chondrites that sheds light on the early solar nebula is their chemical variation. We have seen that the three classes and eight groups of chondrites have different chemical compositions. If meteoriticists are correct that all of these meteorites formed in what is now the asteroid belt, then it is clear that the solar nebula was inhomogeneous on a rather small scale. Is there a pattern to this inhomogeneity that might help us understand the different specific gravities of the terrestrial planets (Figure 10.1)?

There is indeed. It turns out that the chemical differences among chondrites involve three kinds of material: highly refractory elements such as calcium, aluminum, and titanium; iron, nickel, and other elements that enter the crystal structure of metallic iron; and water and other elements that vaporize readily. We call the processes that caused these elements to vary *refractory, siderophile* ("iron-loving"), and *volatile element fractionations*. John Larimer and Edward Anders, who have spent much of their professional lives studying chemical variations in meteorites, examined the first two of these fractionations in 1970. They showed that, to a good approximation, the different refractory element abundances in chondrites can be achieved by starting with the composition of the most sun-like carbonaceous chondrites (CI group) and adding or subtracting material that is chemically similar to CAI's. The chondrule-rich carbonaceous chondrites, for example Allende (CV group), are enriched in such material, while the ordinary (H, L, and LL groups) and enstatite (E) chondrites are deficient in it.

Iron and nickel—the most abundant siderophile elements—likewise follow a rather simple pattern of variation. This pattern becomes obvious if we express the abundances of these elements relative to the abundance of magnesium and correct the average compositions of various chondrite groups for the effects of the refractory element fractionation, which involved magnesium. Figure 10.9, which was constructed from data treated in this way, shows that carbonaceous chondrites, the least intensely metamorphosed enstatite chondrites, and iron-poor (L-group) ordinary chondrites lie close to a common line. As Larimer and Anders observed, it appears that the nickel-iron variation in these chondrites can be explained by a transfer of metal, again starting with material chemically similar to CI chondrites. Since the iron/nickel ratio in the transferred metal is close to the ratio in the sun, it is clear that the metal-silicate fractionation took place before much iron was oxidized or combined with sulfur, that is, at a high temperature. The three groups of ordinary chondrites lie on a different line in Figure 10.9, one whose slope indicates a lower iron/nickel ratio. I have suggested that these meteorites arose through a second transfer of metal, which took place at a lower temperature after iron underwent oxidation and combination with sulfur. According to my reading of the data, chondrites record two episodes of metal-silicate fractionation.

Thus, it appears that the major chemical differences that distinguish the various kinds of chondrites are due to movement of two kinds of material, both of which can be seen among the high-temperature constituents of chondrites: calcium-aluminum-titanium compounds like those found in CAI's, and nickel-iron alloys analogous to meteoritic metal.

Although we are not sure just how these fractionations took place during the formation of chondrites, one of the mobile components—metal—is particularly interesting to planetary scientists because it is about twice as dense as meteoritic silicates. Thus one plausible explanation of the different specific gravities of the terrestrial planets (see

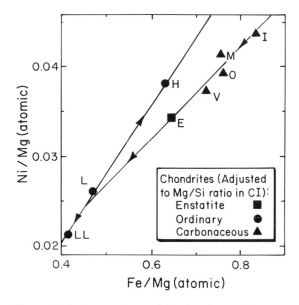

Figure 10.9 Comparison of the iron/magnesium (Fe/Mg) and nickel/magnesium (Ni/Mg) ratios in the eight groups of chondrites. Group mean compositions have been adjusted to restore Mg lost through a refractory element fractionation. Data for the ordinary (L, LL, H) and enstatite (E) chondrites refer to the least metamorphosed members of each group. I, M, O, and V denote the mean compositions of the four groups of carbonaceous chondrites. Arrows show the two inferred metal silicate fractionations discussed in the text.

Figure 10.1) is that they incorporated different proportions of metal: more in Mercury, less in Mars. That possibility will be explored further in the last section of this chapter.

Chondrites also differ in their contents of water, carbon, oxygen, and other volatile constituents: such elements are abundant in carbonaceous chondrites, particularly those (CI) that lack chondrules, and much less abundant in ordinary and enstatite chondrites. This fact suggests a second explanation for the varied specific gravities of the inner planets. Since oxidized iron is less dense than metallic iron and hydrated iron-magnesium silicates are still less dense, it is possible that the terrestrial planets contain the same proportions of the less volatile elements (silicon, iron, magnesium, and so on) but differ in their degrees of oxidation and hydration.

In summary, then, the chondrites tell us that solid nebular materials underwent chemical differentiation before they came together in objects of asteroidal size. That differentiation involved three kinds of material, two of which—metal and volatile constituents—might be responsible for the substantial variations of mean specific gravity that we observe in the terrestrial planets.

Origin of the Solar System

A glance back at the list given at the beginning of the chapter will show that we have found explanations for many properties of the solar system and raised possible explanations for others. The fact that most members of the sun's family rotate and revolve in the same sense and in almost the same plane is an outgrowth of the rotation that the proto-sun acquired from its parent cloud. Exceptions, such as Earth's 23.5-degree axial tilt, are thought to be the results of eccentric impacts late in the growth of planets. Loss of mass during the proto-sun's T-Tauri stage seems to be a satisfactory explanation for the observation that the sun is deficient in angular momentum relative to the planets. Mass ejection may also have cleared the inner solar system of much of its gas before or during the formation of meteorite parent bodies and planets. We have not yet found an explanation for the observation that the terrestrial planets vary in specific gravity and, presumably, chemical composition, but we have found two clues in the metal-silicate and volatile element fractionations that affected chondritic material before it accreted to form parent bodies.

Neither astronomy nor meteoritics has given us much help with the remaining properties on the list: the facts that there are two distinctly different kinds of planets and that the region between them is mass-poor. These observations may make more sense after we try to put together a scenario for the formation of planets.

Building planets. As hard as it is to reconstruct a meteorite parent body from the bits of it that reach the Earth, that task is far simpler than piecing together the sequence of events that led from dust to planets. Though astronomy and meteoritics give us some clues, the gaps that these sciences leave to astrophysics are distressingly wide—wide enough to allow astrophysicists a great deal of leeway in filling them.

In fact, we are not certain just how the process of accretion began. One scenario pictures the early solar nebula as a thin, homogeneous disk of dust and gas and envisions slow, stepwise accretion of particles to form larger and larger bodies in turn. Another proposes that such a disk would be unstable and would break up into a great many small fragments, in much the same way that interstellar clouds break up. In this second scenario the disk fragments, each large enough to have a substantial gravitational field, would collapse and the solids within them would form centrally located protoplanets.

Both these views of the early solar nebula can be defended on physical grounds, and both can produce the terrestrial planets, the former on a time scale of about 100 million years, the latter much more quickly. The former model requires some means of starting the accretion process—getting the first few grains to stick together. The latter needs no such nucleation process, for it produces rather large bodies spontaneously. Against this advantage of the second scenario stands a bigger disadvantage: the concept of gaseous protoplanets implies that most or all of the accretion of the terrestrial planets occurred early, in the presence of solar gas. Since the noble gases trapped in meteorites and those of the Earth have distinctly nonsolar isotopic compositions, it is more likely that the inner planets accreted late, after the proto-sun's T-Tauri ejection drove residual gases from the inner portions of the nebula. For this reason, most recent studies have focused on the first scenario, and so shall I.

Although we believe that the terrestrial planets accreted late, the giant planets most certainly formed early, since they are rich in hydrogen and other gases. Thus Jupiter and Saturn were already in place when the inner planets formed, and their gargantuan gravitational fields had a very strong influence on the behavior of smaller bodies in the evolving solar system.

All students of the early solar nebula agree that it was hot at the center and cooled outward. This follows from the fact that protostars show large and rapid variations of luminosity before they begin to burn hydrogen and settle onto the main sequence. Just how hot the solar nebula was before and during accretion of the inner planets is less clear, but the survival of presolar material in CAI's suggests that it was cooler than about 1500° K (see Figure 10.7). It must have been much cooler in the vicinity of the asteroid belt, cool enough to produce the hydrous silicates in carbonaceous chondrites and perhaps cool enough to form the ice that appears to have been present in their parent bodies. As we shall see, temperature variations in the nebula may have played a large role in the chemical evolution of the inner planets.

Accretion. Most of the recent explanations for accretion of the inner planets picture a thin disk composed of innumerable small particles (dust and, perhaps, chondrules and CAI's) whose total mass equaled the mass of those planets. The particles revolved around the sun in orbits of varied eccentricity and inclination; thus they encountered each other and, presumably, accreted. Jupiter and Saturn were present beyond the region of the inner planets, having grown earlier from the gas-rich nebula. The proto-sun was also present and was heating the disk, but the region of the inner planets was gas-free. Thus the motions of particles in the inner solar system were governed by gravitational effects alone.

How did this swarm of particles evolve into the terrestrial planets? At first, particles with intersecting orbits met and, it is assumed, stuck together. As time passed such encounters produced large bodies at the expense of small ones, eventually yielding a few objects that were massive enough to have substantial gravitational fields. These large bodies, including those that became planets, had a great advantage over the small ones, for they were able to draw in material from the space around them rather than capturing only those particles that happened to strike them. V. S. Safronov, who first discussed these accretion processes in the 1960s, proposed that each proto-planet had its own "feeding zone": a circular tunnel, its diameter determined by the body's gravitational field, all contents of which were incorporated in the proto-planet. This concept of feeding zones has very important implications for chemical variations among the planets.

The mechanism proposed by Safronov for accretion of the terrestrial planets is simple and attractive, but would it produce such objects in a reasonably short period of time? According to calculations made by Stuart Weidenschilling, the answer is yes and no. Using reasonable assumptions, he showed that Mercury, Venus, and the Earth could have grown to their present sizes in about 100 million years, an acceptable figure. Mars, however, with little material dispersed throughout a very large feeding zone, would have taken much longer—about a billion years. Either Mars is much younger than the other terrestrial planets, or its formation involved some process in addition to gravitational accretion.

Weidenschilling prefers the latter explanation, and I agree. The SNC achondrites, which probably sample Mars, record a planetary differentiation that took place 4.5 billion years ago—clearly impossible if the planet formed a billion years later. Weidenschilling, again drawing on an idea set forth first by Safronov, showed that bombardment by objects from the vicinity of Jupiter—perhaps relatives of the present Trojan asteroids—can account for the deficiency of mass in the region of the asteroid belt and Mars without resorting to the idea of a young Mars. He suggested that such objects were also responsible for the intense pounding by meteoroids that the moon experienced about 4 billion years ago. The need to bring in reinforcements to explain Mars makes the Safronov-Weidenschilling hypothesis less tidy than we might wish, but these reinforcements seem to be reasonable, and they explain another of the observations listed at the beginning of the chapter: the deficiency of mass in the region between the Earth and Jupiter. Though Safronov and Weidenschilling admitted that the concept of purely gravitational accretion is probably too simple—a cartoon of the early solar system rather than a portrait—it is a successful cartoon.

Chemical variation. The only problem that still requires an explanation is the observation that the specific gravities of the terrestrial planets decrease outward from the sun. We have seen that either a metal-silicate or a volatile element fractionation might explain this pattern. How might such fractionations be achieved in a system like the one we have been discussing?

As John Lewis demonstrated about a decade ago, a simple and perhaps even correct explanation can be found in the nebular condensation sequence shown in Figure 10.7. If we assume that the nebular temperature was high near Mercury (about 1400° K) and dropped outward, then, neglecting horizontal mixing, solids of different composition would be present at different distances from the sun. At 1400° K, the solid material would consist chiefly of metallic iron, magnesium silicates, and calcium-aluminum-titanium compounds—the only solids that can survive at such a high temperature. A body (Mercury) that formed from those solids would have a high specific gravity because of the high abundance of metallic iron. If Venus formed at a lower temperature, say 900° K, it would also contain metallic iron but would have a larger proportion of magnesium silicates, which condense above 900°. The Earth, still farther from the sun and at a temperature of about 600° K, would contain some metallic iron, plus large proportions of iron sulfide and iron-bearing silicates. Mars, formed at a still lower nebular temperature, would grow from a low-density condensate in which all iron was either oxidized or combined with sulfur. Thus, with a modest amount of cutting and fitting, we can explain the specific gravities of the inner planets by assuming that each one was made from the solids that would be present in nebular gas at a particular temperature. This scheme makes use of *both* of the types of fractionation that we noted in chondrites, for metallic iron is far denser than both silicates and oxidized iron.

This explanation of planetary specific gravities relies on the reasonable assumption that temperature decreased with distance from the proto-sun. It is not necessary to assume that temperature remained constant throughout accretion, nor is that likely: as we have seen, the proto-sun's luminosity probably varied rapidly and widely. All that is required is that the *average* temperature was high at Mercury's solar distance and low at that of Mars.

In its simplest form, Lewis's model implies that all material that formed at a particular distance from the sun accreted in the same place. This requirement is neither necessary nor realistic. We know that some mixing must have taken place among the planets, since the Earth contains a great many constituents, most notably carbon and water, that should have remained in the nebular gas at 600° K. How did such mixing occur?

Safronov realized that after a planet exhausted its own feeding zone, it could continue to grow only by adding material that wandered in from other zones. Though Safronov felt that a planet would acquire very little of this exotic material, William Hartmann has shown that mixing among the planetary feeding zones was probably extensive. The Earth, for example, obtained almost 40 percent of its mass from outside its zone. The moon imported still more—about 60 percent—and got most of it from the vicinity of Venus. Hartmann suggests that this exotic material may be responsible for the moon's low specific gravity and its deficiency in volatile constituents.

Although modern explanations for the varied specific gravities of the inner planets differ considerably in detail, most of them are variations on the same theme: The solids that accreted to form the planets varied chemically in response to a temperature gradient. Much of the material that formed each planet came from nearby to give it its distinctive composition and specific gravity. Throughout accretion, but particularly in its later stages, each planet also acquired material from more distant sources—enough to blur, but not erase, the chemical distinctions that make the four terrestrial planets and the moon five quite different worlds.

11

Meteorites and Life on Earth

We saw in Chapter 1 that meteorites have played only a minor role in the chemical and physical evolution of the Earth during the last 4 billion years. The flux of extraterrestrial material is too small to change significantly the composition of our planet's crust, and impacts by giant meteoroids are too infrequent to compete with other processes—sea-floor spreading, mountain building, volcanism, erosion—that shape its surface. In this sense, meteorites and their impact phenomena are just footnotes to geological history.

They may have played a much larger role in the Earth's biological history. During the last quarter-century, two exciting ideas have emerged concerning possible relationships between meteorites and life on Earth. One is that life existed at the time and place where carbonaceous chondrites formed, and those meteorites brought it to our planet. The other is that impacts by huge meteoroids—asteroids or comet nuclei—have punctuated organic evolution, extinguishing some life forms and allowing others to go forward and prosper.

The first idea, which had a small but vocal following during the early 1960s, sprang from two observations: that some carbonaceous chondrites contain a remarkably varied suite of organic (hydrogen-carbon-nitrogen-oxygen) compounds, some of which resemble products of living tissue; and that the same meteorites contain tiny objects—dubbed "organized elements"—that resemble microscopic fossils. The first of these observations has been confirmed and extended in recent years through meticulous study of additional carbonaceous chondrites, including some of the less weathered specimens from Antarctica. These meteorites do indeed contain complex organic compounds, including amino acids—the building blocks of proteins.

The second observation has fared less well. Although a great many "organized elements" were described in the early 1960s, all proved to be something other than fossils: terrestrial contaminants in some cases (for example, pollen), oddly formed minerals in others. Though we cannot yet prove that there are no fossils in carbonaceous chondrites—to prove there are no green cats, one must examine all cats— there is now no positive evidence for such objects. Nor does the existence of complex organic compounds in meteorites testify to biological activity. The compounds observed in carbonaceous chondrites have been duplicated in laboratory experiments by purely chemical processes. Thus we are forced to conclude that the idea that life was present in the meteorite parent bodies and came to Earth via meteorites has no factual basis; it was one of the many blind alleys of science.

The second idea—that meteoroidal impacts have influenced the course of evolution—has fared far better. Once regarded as speculation or worse, it now has considerable support and is one of the most exciting hypotheses that has swept through the Earth sciences since sea-floor spreading appeared 25 years ago.

Paleontologists have known for many years that the slow pace of organic evolution that they see recorded in fossils is broken at many points by abrupt changes. After millions of years of gradual replacement of old organisms by new ones, many life forms leave the stage in a geologically short period of time, to be replaced by a very different cast of characters. This phenomenon, which paleontologists call *mass extinction,* has occurred at least 15 times in the last 600 million years (Figure 11.1).

One of the most impressive episodes of mass extinction took place at the end of the Cretaceous period and included the disappearance of the dinosaurs and other giant reptiles. These animals, which had evolved from humble late Paleozoic ancestors, all but ruled the Earth during the Mesozoic era. Late in that era, horned, tank-like ceratopsians (for example, the familiar dinosaur *Triceratops*) and duck-billed hadrosaurians roamed the lands; marine lizards (mosasaurs) prowled the seas; and pterosaurs patrolled the skies. Then, as the Cretaceous period gave way to the Tertiary, these giant reptiles vanished, survived by just a few modest lizards and crocodiles and—according to some paleontologists—the birds.

Though many other kinds of animals and many plants also disappeared at the end of the Cretaceous period, the demise of the di-

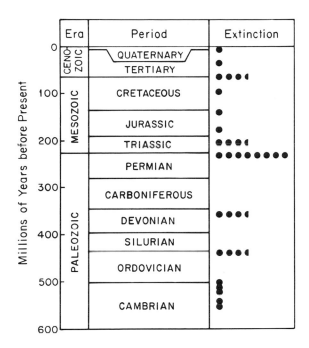

Figure 11.1 Geological time scale for the last 600 million years, showing times of mass extinction. The number of dots reflects the relative intensity of each extinction. (Data are from Sepkoski, 1982.)

nosaurs is particularly intriguing. What killed these formidable animals? Did the climate become too hot or too cold for them? Did the shallow seaways along which many of them lived dry up, perhaps because the continents rose? Or did tiny, primitive mammals, waiting restlessly for their time in the evolutionary spotlight, extinguish their huge competitors by eating their eggs?

None of these ideas is a completely satisfactory way to explain the dinosaurs' disappearance, and none explains why many other kinds of life forms vanished during the same brief period of time. Since no conventional explanation seemed adequate, two bold scientists—Digby McLaren in 1969 and Harold Urey in 1973—proposed a most unconventional one: that the mass extinction at the end of the Mesozoic era was caused by a collision between the Earth and one of its neighbors in space, an asteroid or a comet. This suggestion caused a brief sensation in the scientific community, but most geologists greeted it with polite silence. There was no positive evidence for a major impact at the end of the Cretaceous period. Moreover, the impact scenario seemed to violate a basic tenet of geology: that the Earth and its inhabitants have evolved gradually, through the action of processes that we can still observe today.

The impact interpretation might well have languished as fanciful speculation but for the efforts of four imaginative scientists from the

University of California at Berkeley. In 1978 Luis Alvarez, Walter Alvarez, Frank Asaro, and Helen Michel made a startling discovery: a thin layer of clay that lies at the Cretaceous-Tertiary boundary in Gubbio, Italy, is exceptionally rich in iridium, an element that is sparse in rocks of the Earth's crust but relatively abundant in meteorites. The Berkeley team proposed that a small asteroid, about 10 kilometers (6 miles) in diameter, struck the Earth 65 million years ago, depositing the iridium-rich layer and, as a result of atmospheric effects related to its impact, extinguishing many forms of life.

The impact hypothesis for the Cretaceous-Tertiary mass extinction has shaken the foundations of geology and opened our eyes to the possibility of still more intriguing relationships between the Earth and its neighbors in space. In this chapter we shall see where that hypothesis stands today, where it is heading, and what it means to our view of the Earth.

The Cretaceous-Tertiary Mass Extinction

Of the fifteen mass extinctions shown in Figure 11.1, most were modest affairs that involved only a slight increase in the rate of extinctions. However, four were dramatic and one was spectacular. Curiously, the most severe mass extinction was not the one that closed the Mesozoic era, but one that occurred late in the preceding Paleozoic era. During the waning stages of the Permian period, more than 50 percent of the families of marine life and perhaps as many as 96 percent of their species became extinct in a biological collapse so devastating that one geologist called it "the time of the great dying."

In contrast to the abrupt Cretaceous-Tertiary event, the extinction that closed the Permian period was quite leisurely: a steep decline in the diversity of life forms over a period of 15 million years. It also took place at a time when the Earth's surface was undergoing profound changes brought about by the breakup of a gigantic land mass whose fragments would form most of the modern continents. Paleontologists believe that this reshuffling of the continents changed patterns of circulation in the oceans and atmosphere, and the environmental stresses caused by those changes were responsible in turn for the high mortality of life forms at the end of the Permian period.

Though the Cretaceous-Tertiary mass extinction was less dramatic than that in the Permian period, it eliminated about 15 percent of the existing families of marine animals and larger percentages of genera and species for these and many other forms of life (see Figure 11.1). Moreover, this extinction was surprisingly selective: it affected marine animals far more severely than those based on land, and among the former, swimmers (for example, the nautilus-like ammonites and the squid-like belemnites) more than bottom-dwellers (Figure 11.2). In some cases, two animals with very similar habits and habitats fared differently; thus marine lizards perished, but marine crocodiles survived.

A very curious feature of the Cretaceous-Tertiary extinction, and one that is hard to explain, is that it had drastically different effects on plants in the temperate mid-latitudes and in the tropics. More than 70 percent of the plant genera that occupied the mid-latitudes during the Cretaceous period were gone by the beginning of the Tertiary period. In contrast, the number of tropical genera *increased* by 50 percent across the Cretaceous-Tertiary boundary (Hickey, 1981).

Most paleontologists agree that the pace of organic evolution quickened at the end of the Cretaceous period, but they differ widely on the nature of the change. Did previously healthy populations of organisms disappear abruptly 65 million years ago, or were the life

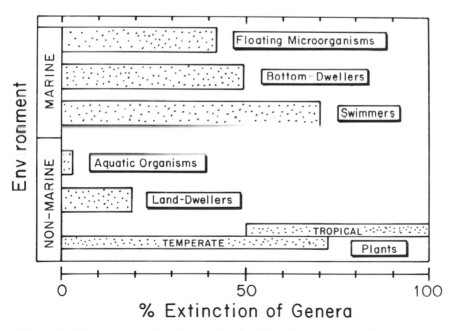

Figure 11.2 Percentages of extinction for families of organisms at the Cretaceous-Tertiary boundary. Note that the number of tropical plant families *increased* by 50 percent across the boundary. (Data are from Russell, 1979, and Hickey, 1981.)

forms that vanished already in decline long before the end of the period? Should we picture the Cretaceous-Tertiary mass extinction as a sharp inflection in the evolutionary curve—a cliff—or just a steep slope?

The fossil record suggests different histories for different organisms. Coccoliths and foraminifera, two groups of floating marine microorganisms that build shells of calcium carbonate, were almost completely extinguished in a time interval that has been estimated by different researchers as between 50 and 10,000 years. Their extinction was, by any reasonable definition, catastrophic (Tappan, 1982). On the other hand, most paleontologists believe that the dinosaurs had already dwindled to a few geographically isolated species by the end of the Cretaceous period, and many infer a similar decline for ammonites.

It is possible that the historical contrast between marine plankton and dinosaurs is more apparent than real, for the fossil record is much less complete for the latter than for the former. Dinosaur bones are preserved in relatively few places, and they are difficult and costly to recover. Some paleontologists feel that the record for dinosaurs is too sparse to justify the conclusion that they were waning well before the end of the Cretaceous period (Russell, 1982). This argument is less persuasive for ammonites, whose fossils are widespread in space and time, but here too paleontologists differ on how, and how rapidly, these marine animals became extinct.

Evidence that some life forms were declining well before the end of the Cretaceous period seems at first glance to testify against a catastrophe at the end of that period. However, decline, extinction, and replacement have occurred throughout the history of life: if a flood ravaged a town, it would kill both healthy men and women and people who, by virtue of age or illness, would have died shortly anyway. The impressive feature of the Cretaceous-Tertiary mass extinction is that both declining life forms (dinosaurs, ammonites) and forms that were apparently robust (coccoliths, foraminifera) vanished at nearly the same point in time. In contrast with the Permian event that preceded it, the Cretaceous-Tertiary mass extinction was abrupt.

Testing the Impact Hypothesis

The impact hypothesis meets the requirements of suddenness and brevity far more readily than any other explanation for the late Mesozoic mass extinction that has been proposed to date. The Earth's recent experience with continental glaciation shows us that major climatic changes take place on time scales of thousands to millions of years. Dramatic shifts of the continents and profound changes in oceanic circulation are at least that leisurely. Such conventional geological mechanisms can account for the prolonged Permian mass extinction, but they seem quite inadequate to explain the sudden changes at the end of the Cretaceous period.

To evaluate the impact hypothesis, we have to ask three questions. First, is it plausible that a small asteroid or comet struck our planet 65 million years ago? Second, can we prove beyond reasonable doubt that one did? Finally, would a major impact produce secondary effects that could destroy many forms of life but leave others untouched?

Plausibility. We can estimate the probability of major impacts on Earth by studying the distribution of large objects near the Earth (see Chapter 1) and by examining our planet's impact record. There are now about 1,000 Earth-crossing asteroids with diameters of at least 1 kilometer (0.6 mile). Since their frequency falls off with increasing size, small asteroids are far more likely to strike the Earth than large ones. Nevertheless, recent calculations suggest that a 10-kilometer asteroid—the size suggested by the Berkeley group—should land every 40 million years or so. Impacts by comet nuclei of the same size should be somewhat less frequent.

Another way to investigate the likelihood of a major impact 65 million years ago is to examine the distribution of terrestrial meteorite craters. There are about 100 craters whose impact origin has been confirmed by the discovery of nearby meteorites, shock features in the surrounding rocks, or both. Figure 11.3 shows their geographical distribution, Figure 11.4 the sizes of those craters whose ages are known. Terrestrial impact craters range in diameter from a few meters to 140 kilometers (85 miles) and in condition from pristine (for example, Meteor Crater; see photograph preceding p. 1) to very deeply eroded (for example, Manicouagan, Quebec; Figure 11.5). All craters discovered to date are on land, and most are in areas that have not been uplifted or eroded for a very long time. A great many occur on the ancient metamorphic-igneous shields that cover much of northern North America, Europe, and Asia.

All but the biggest craters on Earth are less than a billion years old. The exceptions, Vredefort in South Africa and Sudbury, Ontario, are 85 miles wide and 1,970 million and 1,840 million years old, respectively. The facts that there are few very old craters and that these craters are typically large reflect the ease with which these features are destroyed by erosion. In the case of the Manicouagan crater (Figure

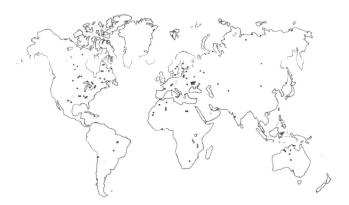

Figure 11.3 Geographical distribution of verified impact craters. Open symbols mark craters with associated meteorites. (From Grieve, 1982, reproduced by permission of the author and the Geological Society of America.)

11.5), we see only its lower portion. If Meteor Crater were worn away to the same depth, it would vanish altogether.

In places where terrestrial craters are fresh enough for detailed study, we find that many contain shock-melted material and most are rimmed by shattered and pulverized material ejected during impact. Although most of this material lies close to its source, some of it traveled for remarkably long distances. For example, some craters are associated with *tektites,* scattered beads and blobs of silica-rich, impact-generated glass, some of which show the effects of aerodynamic shaping during their flight through the atmosphere (Table 11.1; Figure 11.6). The tektite strewnfields, one of which may extend halfway around the Earth (Figure 11.7), testify to very wide dispersal of coarse material from craters of even moderate size. Fine material travels still further. Dust from the Tunguska, Siberia, fall in 1908 circled the globe and brightened the night sky over much of Europe and Asia.

I shall return to the known and predicted effects of major impacts later in this chapter; for now it suffices to note that both the size distribution of Earth-crossing asteroids and comets and the frequency of large terrestrial craters make a major impact at the end of the Cretaceous period plausible.

Evidence of a terminal Cretaceous impact. The strongest evidence that a major impact occurred 65 million years ago would, of course, be a large crater of that age. Unfortunately, we know of no such crater. The few craters that are about the right age are too small to account for major, worldwide climatic effects (see Figure 11.4).

This result is disappointing, but it does not rule out a major terminal Cretaceous impact. There is a very good chance (roughly three out of four) that the projectile landed in the ocean. If a suboceanic crater survived—much of the ocean floor has slid beneath the continents during the last 65 million years—it would be hard to find without very detailed bathymetric and geophysical studies. It is also possible that the Cretaceous-Tertiary event involved a swarm of small asteroids or comets rather than one big one. Finally, as the Tunguska fall illustrates, an asteroid or, more probably, a comet head might explode in the atmosphere, scattering material far and wide but leaving no crater. In the light of these possibilities, failure to locate a suitable crater causes the proponents of a late-Cretaceous impact little concern.

More impressive to the advocates of the impact hypothesis are the results of further studies of sediments at the Cretaceous-Tertiary boundary. Since Alvarez and his colleagues made their original discovery at Gubbio, enhancements of iridium and other meteorite-correlated elements have been found in almost 50 other samples from the boundary, including some that formed on the continents. It is now clear that the boundary anomaly is worldwide and involves many elements other than iridium.

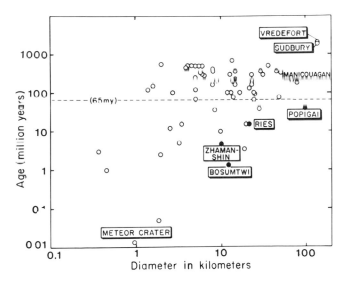

Figure 11.4 Sizes and ages of dated terrestrial craters, based on data collected by Grieve (1982). Craters that are, or may be, associated with tektite strewnfields are indicated by filled circles; all craters mentioned in the text are labeled.

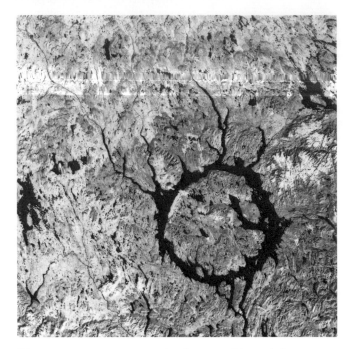

Figure 11.5 Manicouagan, Quebec, an old, eroded crater. The visible crater, outlined by lakes, is 70 km (42 miles) wide, but fractures related to it extend outward for another 30 km. (NASA ERTS photograph, no. E-1438-15024-6.)

The worldwide extent of this anomaly has led most geologists to acknowledge that a major influx of extraterrestrial material took place 65 million years ago, but a few have held back to consider other explanations. Might the elements that are concentrated at the boundary be volcanic? If they are meteoritic, do they necessarily reflect just one episode of impact, or may they be material that was originally dispersed in the limestones below the boundary and concentrated by solution of those limestones?

Two observations make us quite certain that the concentration of iridium and related elements in the boundary clay is not due to volcanism: one is the fact that the boundary clay contains mineral grains that bear unequivocal evidence of impact shock; the other is that osmium, an element that is closely related to iridium, has isotopic ratios in the clay that closely resemble the ratios observed in meteorites but differ from those found in terrestrial rocks. It is hard to avoid the conclusion that the anomalous elements in the boundary clay came from beyond the Earth. The second alternative is harder to dismiss out of hand, for solution of the limestone below the boundary would indeed concentrate the meteoritic material within it. However, the properties of the limestones beneath the boundary seem to be inconsistent with extensive solution. Moreover, it is hard to see how solution could have gone on all over the world at the same time. This alternative too strains belief.

However they view the relationship between impact and mass extinctions, most geologists now accept the Berkeley group's suggestion that a small asteroid, a comet, or a swarm of such objects struck the Earth at the end of the Cretaceous period.

Mechanisms of extinction. Two questions remain to be resolved: what effects would accompany and follow an impact by a large extraterrestrial object, and can these effects account for the observed changes in terrestrial life at the end of the Cretaceous period? Because research on these questions is still in full flood, what follows is more a progress report than a conclusive statement.

The extraterrestrial body that Alvarez and his co-workers envisioned (6 miles in diameter, with a specific gravity of 3 and traveling at 15 miles per second) would hit the Earth with the energy of 100 million tons of TNT—much more energy than is held in all of the world's nuclear arsenals. If it struck land, it would produce a crater between 60 and 75 miles wide. Most of the material ejected from that crater would land near it, but a lot—perhaps twice the mass of the body—would be driven upward into the atmosphere. An impact in water would be even more dramatic. It would excavate a slightly wider crater and throw much more material into the atmosphere: salt steam as well as pulverized, melted, and vaporized rock. Such an impact would breach the Earth's second layer, or mantle, which lies

Table 11.1 Tektite strewnfields and impact craters to which each is or may be related

Strewnfield	Age (millions of years)	Crater	Extinction
Australasian	0.7	Zhamanshin, USSR (?)	No
Ivory Coast	1.0	Bosumtwi, Ghana	No
Czechoslovakian	14.7	Ries, West Germany	No
North American	34.4	Popigai, USSR (?)	Possible

Note: The North American tektites and their microscopic equivalents are correlated with an iridium anomaly and, perhaps, a modest mass extinction late in the Eocene epoch of the Tertiary period.

Figure 11.6 (*top, middle*) Different views of a one-inch Australasian tektite, showing spherical form modified by an ablation cap. (*bottom*) Two larger but less well-formed tektites from the Philippines. (Photographs by Rota, courtesy of the Department of Library Services, American Museum of Natural History; negative nos. 2A-8889, 2A-8890, 2A-7725.)

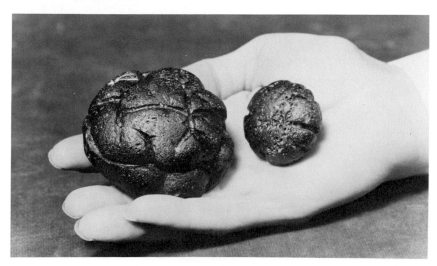

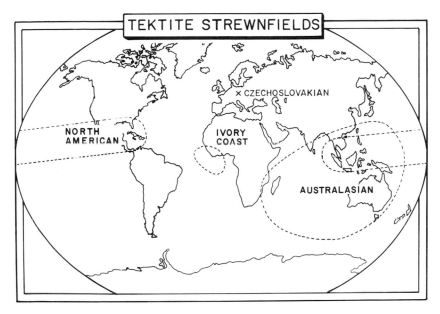

Figure 11.7 Locations of major tektite strewnfields. (Generalized from Glass, 1982, and reproduced by permission of the author and the Geological Society of America.)

at shallow depths beneath the oceans. It would also generate sea waves, or *tsunamis,* which would dwarf those produced by earthquakes and would swiftly flood the world's coasts.

Just how such impacts would affect terrestrial life would depend strongly on the fate of impact ejecta in the atmosphere: how fast they settled and how they changed the atmosphere's chemistry. The best estimates to date suggest that the dust cloud would plunge the Earth's surface into total darkness for 2 to 12 months. With sunlight blocked for several months to a year, the Earth's surface would quickly grow cold. Though the oceans would cool by just a few degrees, subfreezing temperatures would sweep across the continents and snow would follow, accumulating to depths of perhaps 15 to 20 feet. Even when the sun emerged from the impact pall, this snow cover would reflect its light efficiently to slow the return of warmer temperatures. For months or even years, the Earth would be very bright but very cold.

Though the scenario described above implies that only ejecta from the impact would have been available to darken the sky, Wendy Wolbach and her co-workers at the University of Chicago (1985) have proposed that smoke was present as well. Abundant carbon in clay samples from the Cretaceous-Tertiary boundary suggests that the impact kindled immense fires, which decimated plant life on the continents and intensified and prolonged the postimpact pall.

The lethal gloom and chill after a major impact would be accompanied by chemical changes in the atmosphere. When the Tunguska meteoroid exploded over Siberia, it produced nitrogen oxides that destroyed 35 to 45 percent of the Earth's ozone layer. These compounds would form on a far grander scale after the impact of a small asteroid or comet. Among their effects would be the formation of nitric acid, which would rain down on and poison the upper layers of the oceans, destroying the delicate chemical balance that allows many marine animals and plants to build calcium carbonate shells.

We are reasonably sure of these immediate consequences of a major impact. We are much less sure just how, and how fast, the Earth would recover from them. If snowfields were widespread on the continents, surface temperatures would moderate very slowly. Indeed, some atmospheric scientists have suggested that a slightly more vigorous impact 65 million years ago might have produced enough snow to leave our planet's surface completely and forever frozen. On the other hand, the aftermath of refrigeration might well be a period of abnormally high temperatures, since wholesale destruction of green plants and solution of carbonates in the soured oceans would charge the atmosphere with carbon dioxide. This carbon dioxide would produce a greenhouse effect similar to, but much more intense than, the one that man has created by burning fossil fuels.

Clearly, the aftermath of a large impact offers a broad menu of lethal possibilities: darkness; intense cold, perhaps followed by heat; an acid atmosphere and oceans.* How do these possibilities square with the observed patterns of extinction at the end of the Cretaceous period? On the whole, quite well. On land, darkness, cold, and perhaps fires would quickly destroy green plants, starving herbivorous animals and the carnivores that depend on them. Large, cold-blooded animals with immense food requirements would be particularly vulnerable to the effects of darkness and cold. Small, warm-blooded animals, especially nocturnal species and those accustomed to winter hibernation, might, on the other hand, survive. Thus dinosaurs would

---

*The reader may be struck by the resemblance between the results of a major impact and the disasters that man has visited, or threatens to visit, on the Earth and its inhabitants. Acid rain, destruction of the ozone layer, the greenhouse effect, and the natural equivalent of nuclear winter are, it seems, nothing new. Man has developed some small proficiency in environmental vandalism, but he did not invent it.

almost certainly be destroyed, but small, burrowing mammals might survive.

In the oceans too, darkness would kill green plants and the animals that feed on them, sending a wave of starvation up the food chain. Acidity would also destroy those animals (for example, coccoliths and foraminifera) and plants that have calcium carbonate shells. This effect, like that of darkness, would be most lethal to organisms that live near the surface of the ocean, such as ammonites. Bottom-dwellers, accustomed to eternal darkness and cold, would be more likely to survive the postimpact pall. They would also be protected against the effects of acid rain, which would be mixed with and diluted by normal seawater by the time it reached them.

Although the impact scenario can explain most of the broad features of the Cretaceous-Tertiary mass extinction, many serious problems remain unsolved. The survival pattern for plants is a good example. If darkness and cold destroyed plants, we might expect the hardier high-latitude species to fare better than plants in the warm, light tropics. In fact, the former were decimated at the end of the Cretaceous, while the latter flourished. There may be ways to solve this puzzle. Presumably, the "meteoroid winter" was longer and colder in the high latitudes than it was closer to the Equator. Perhaps the longer winter destroyed the seeds of temperate plants, but those of tropical species survived a shorter, milder refrigeration to germinate when light and heat returned to the Earth. Or perhaps it was unaccustomed heat—the greenhouse effect—that winnowed out plant species, allowing tropical forms to prosper while temperate forms withered.

Status of the impact hypothesis. The results of a poll taken in 1984 of 172 American paleontologists showed that 55 percent accepted the idea of a major impact at the end of the Cretaceous period, but only 16 percent thought that it was responsible for mass extinction (Hoffman and Nitecki, 1985). Although consensus is a poor basis for scientific decisions—in 1800, the French Academy of Sciences would have voted solidly against an extraterrestrial origin for meteorites; in 1950, most members of the Geological Society of America would have rejected continental drift—this poll is significant, for it shows scientists doing what they should: reserving judgment on a new hypothesis until the evidence for it becomes irrefutable.

The evidence that an asteroid or comet turned the stream of evolution 65 million years ago is far from complete, but the idea has injected new excitement into geology and stimulated a wave of research that has yet to crest. Even if the impact hypothesis proves to be wrong (which in my opinion is unlikely), the new knowledge gained from efforts to test it will be worth their cost.

Other Mass Extinctions

I have discussed just two of the fifteen mass extinctions shown in Figure 11.1 and have shown that one of these, the one that ended the Paleozoic era, was almost certainly caused by wholly terrestrial processes. What of the others? Most extinction events are less well documented than those described here, but some modest mass extinctions recorded in Cambrian sediments have been examined with care. The rocks in which they occur appear to contain no meteoritic material, and it seems likely that those events, like that in the Permian period, were caused by terrestrial processes.

To date, only an event near the end of the Eocene epoch of the Tertiary period shows evidence of a connection between impact and mass extinction, and that evidence is in dispute. A deep-sea sediment core from the Caribbean contains a layer that is enriched in iridium. It lies just above a concentration of microtektites, which appear on the basis of fission-track dating to be 34 million years old. The iridium-rich layer also appears to be correlated with the disappearance of five species of radiolarians. Though the pieces seem to fit together nicely, some researchers have disputed the uniqueness and age of the microtektite horizon and its time correlation with the extinctions. Since this issue remains unresolved, I have described the late Eocene event as "possible" in Table 11.1.

While the search for further evidence for (or against) impact-related mass extinctions goes on, the new dialogue between paleontologists and planetary scientists has produced other exciting results. One is a hint that clusters of asteroids, or more likely comets, strike the Earth at intervals of about 26 million years. Some paleontologists think that they see a similar periodic variation in extinction rates, and many researchers are now striving to verify (or falsify) these patterns and explain them.

It is much too early to tell whether the idea of periodic impacts by extraterrestrial objects will blossom into a still grander view of the Earth's relation to other members of the sun's family or will wither before a fiery blast of new data, but it shows that the romance between geology and planetary astronomy that began with the manned space program is far from over. Indeed, it has barely begun.

References

Index

References

Bjorkman, J. K. 1973. Meteors and meteorites in the ancient Near East. *Meteoritics* 8:91–130.

Chapman, C. R. 1976. Asteroids as meteorite parent bodies: The astronomical perspective. *Geochimica et Cosmochimica Acta* 40:701–719.

Clayton, R. N., and T. Mayeda. 1978. Genetic relations between iron and stony meteorites. *Earth and Planetary Science Letters* 40:168–174.

——— 1983. Oxygen isotopes in eucrites, shergottites, nakhlites and chassignites. *Earth and Planetary Science Letters* 62:1–6.

Dodd, R. T. 1981. *Meteorites: A Petrologic-Chemical Synthesis.* Cambridge: Cambridge University Press.

Dodd, R. T., E. J. Olsen, and R. S. Clarke, Jr. 1985. The Bloomington breccia and its shock melt glasses. *Meteoritics* 20:575–581.

Glass, B. P. 1982. Possible correlations between tektite events and climatic changes? In Special Paper 190, *Geological Implications of Impacts of Large Asteroids and Comets on the Earth,* ed. L. T. Silver and P. H. Schultz. Boulder, Colo.: Geological Society of America, pp. 251–256.

Graham, A. L., A. W. R. Bevan, and R. Hutchison. 1985. *Catalogue of Meteorites,* 4th ed. London: British Museum (Natural History), and Tucson: University of Arizona Press.

Grieve, R. A. F. 1982. The record of impact on Earth: Implications for a major Cretaceous/Tertiary impact event. In Special Paper 190, *Geological Implications of Impacts of Large Asteroids and Comets on the Earth,* ed. L. T. Silver and P. H. Schultz. Boulder, Colo.: Geological Society of America, pp. 25–38.

Hey, M. H. 1966. *Catalogue of Meteorites,* 3rd ed. London: British Museum.

Hickey, L. J. 1981. Land plant evidence compatible with gradual, not catastrophic change at the end of the Cretaceous. *Nature* 292:529–531.

Hoffman, A., and M. H. Nitecki. 1985. Reception of the asteroid hypothesis of terminal Cretaceous extinctions. *Geology* 13:884–887.

Hughes, D. W. 1981. Meteorite falls and finds: Some statistics. *Meteoritics* 16:269–281.

Hutchison, R., A. W. R. Bevan, and J. M. Hall. 1977. *Appendix to the Catalogue of Meteorites.* London: British Museum.

189

Lovering, J. F. 1957. Differentiation in the iron-nickel core of a parent meteorite body. *Geochimica et Cosmochimica Acta* 12:238–252.

McSween, H. Y., Jr. 1984. SNC meteorites: Are they Martian rocks? *Geology* 12:3–6.

Richardson, S. M. 1978. Vein formation in the C1 carbonaceous chondrites. *Meteoritics* 13:141–159.

Romig, A. D., Jr. and J. I. Goldstein. 1979. Determination of the Fe-Ni and Fe-Ni-P phase diagrams at low temperatures (700-300 C). *Metallurgical Transactions Acta* 11A:1151–9.

Russell, D. A. 1979. The enigma of the extinction of the dinosaurs. *Annual Reviews of Earth and Planetary Sciences* 7:163–182.

——— 1982. A paleontological consensus on the extinction of the dinosaurs? In Special Paper 190, *Geological Implications of Impacts of Large Asteroids and Comets on the Earth*, ed. L. T. Silver and P. H. Schultz. Boulder, Colo.: Geological Society of America, pp. 401–405.

Scott, E. R. D., and J. T. Wasson. 1975. Classification and properties of iron meteorites. *Reviews of Geophysics and Space Physics* 13:527–546.

Sears, D. W. 1975. Sketches in the history of meteoritics 1: The birth of the science. *Meteoritics* 10:215–225.

Sepkoski, J. J., Jr. 1982. Mass extinctions in the Phanerozoic oceans: A review. In Special Paper 190, *Geological Implications of Impacts of Large Asteroids and Comets on the Earth*, ed. L. T. Silver and P. H. Schultz. Boulder, Colo.: Geological Society of America, pp. 283–290.

Stafford, E. P. 1980. A four-year fight to bring home a giant meteorite. *Smithsonian* 11 (no. 2):133–148.

Tappan, H. 1982. Extinction or survival: Selectivity and causes of Phanerozoic crises. In Special Paper 190, *Geological Implications of Impacts of Large Asteroids and Comets on the Earth*, ed. L. T. Silver and P. H. Schultz. Boulder, Colo.: Geological Society of America, pp. 265–276.

Urey, H. C., and H. Craig. 1953. The composition of stone meteorites and the origin of the meteorites. *Geochimica et Cosmochimica Acta* 4:36–82.

Wasson, J. T. 1974. *Meteorites: Classification and Properties*. New York: Springer-Verlag.

Wolbach, W. S., R. S. Lewis, and E. Anders. 1985. Cretaceous extinctions: Evidence for wildfires and search for meteoritic material. *Science* 230:167–170.

Wood, J. A. 1979. *The Solar System*. Englewood Cliffs, N.J.: Prentice-Hall.

Index